ENVIRONMENTAL SCIENCE, ENGINEERING AND TECHNOLOGY

AF783728

REMOTE SENSING OF SOLAR INFLUENCE ON ANTARCTIC TERRESTRIAL CLIMATE FROM A GPS PERSPECTIVE

Environmental Science, Engineering and Technology

Additional books in this series can be found on Nova's website under the Series tab.

ENVIRONMENTAL SCIENCE, ENGINEERING AND TECHNOLOGY

REMOTE SENSING OF SOLAR INFLUENCE ON ANTARCTIC TERRESTRIAL CLIMATE FROM A GPS PERSPECTIVE

WAYAN SUPARTA

Nova Science Publishers, Inc.
New York

Copyright © 2010 by Nova Science Publishers, Inc.

All rights reserved. No part of this book may be reproduced, stored in a retrieval system or transmitted in any form or by any means: electronic, electrostatic, magnetic, tape, mechanical photocopying, recording or otherwise without the written permission of the Publisher.

For permission to use material from this book please contact us:
Telephone 631-231-7269; Fax 631-231-8175
Web Site: http://www.novapublishers.com

NOTICE TO THE READER

The Publisher has taken reasonable care in the preparation of this book, but makes no expressed or implied warranty of any kind and assumes no responsibility for any errors or omissions. No liability is assumed for incidental or consequential damages in connection with or arising out of information contained in this book. The Publisher shall not be liable for any special, consequential, or exemplary damages resulting, in whole or in part, from the readers' use of, or reliance upon, this material.

Independent verification should be sought for any data, advice or recommendations contained in this book. In addition, no responsibility is assumed by the publisher for any injury and/or damage to persons or property arising from any methods, products, instructions, ideas or otherwise contained in this publication.

This publication is designed to provide accurate and authoritative information with regard to the subject matter covered herein. It is sold with the clear understanding that the Publisher is not engaged in rendering legal or any other professional services. If legal or any other expert assistance is required, the services of a competent person should be sought. FROM A DECLARATION OF PARTICIPANTS JOINTLY ADOPTED BY A COMMITTEE OF THE AMERICAN BAR ASSOCIATION AND A COMMITTEE OF PUBLISHERS.

LIBRARY OF CONGRESS CATALOGING-IN-PUBLICATION DATA

Available upon Request
ISBN: 978-1-61761-317-3

Published by Nova Science Publishers, Inc. † New York

CONTENTS

PREFACE

Water vapor is a key in the hydrological cycle and the main driver of atmospheric events. Remote sensing of water vapor using the Global Positioning System (GPS) is essential for weather and climate research. Advances in GPS application has allowed the measurements of ionospheric total electron content (TEC) and precipitable water vapor (PWV) in all weather conditions and on a global scale with fine temporal and spatial resolution. The two quantities are represents the upper and the lower levels of the atmosphere which can be used for study the recent solar activity-weather/climate relationships. Using TEC as a measure of solar activity and PWV as the terrestrial response, a new approach for determining and quantifying the solar influence on PWV by indirectly correlating the PWV and TEC variations on a time basis is proposed. Solar related events influence on terrestrial climate during transient events associated with the intense of geomagnetic storms is observed and investigated. In the analysis, the morphology of solar modulated geomagnetic activity towards induced atmospheric water vapor to understand the possible physical mechanism of how solar activity influences terrestrial weather/climate changes at the current state of the art of GPS technique are described and discussed.

Chapter 1

INTRODUCTION

Water vapor content and their circulation on the globe was a growing concern in many parts of the world due to its contributed to the many physical processes in the Earth's atmosphere such as global warming and climate change developments. Global warming issues were take place, since an increase in atmospheric greenhouse gases due to human activity. Increased global surface temperature of 0.74 °C (0.56 °C to 0.92 °C) during the 100 years ending in 2005 is expected due to increased in anthropogenic greenhouse gases concentrations [1]. Although water vapor constitutes only a small fraction of all the atmospheric gases, it has crucial roles in affecting both climate and weather. As a source of precipitation, water vapor is the biggest contributor to the 'natural greenhouse effect' and this 'positive feedback' makes it plays a critical role in the global climate system. Water vapor changes rapidly and has irregular variability in both temporal and spatial scales, from micro to global meteorology [2]. For example, water vapor plays a major role in the dynamics of atmospheric circulation which is fundamental to the transfer of energy and in the formation and propagation of weather [3]. Water vapor also the source of precipitation and its latent heat is a critical ingredient in the dynamics of most major weather events. The role of water vapor is not only restricted to absorbing and radiating energy from the sun [4], but also includes the effect it has on the formation of clouds and aerosols, and the chemistry of the lower atmosphere [5]. In addition, the distribution of water vapor is closely associated with the clouds, rainfall, convection and storms [5, 6, 7]. It also has a significant attenuation effect at millimeter wavelength and has been positively correlated with cloud attenuation [8, 9]. On the other hand, water vapor is one climate variable and is a key in the hydrological cycle and the main driver of atmospheric events. Besides its importance to atmospheric processes over a wide range of spatial and temporal scales, it is essentially important for numerical weather prediction (NWP). Improved knowledge of its distribution is needed for a variety of atmospheric research applications and for improved weather forecasting [10]. The comprehensive understanding of its distribution and mechanisms will have a major impact on a better planning of resources as well as the key to a better understanding of global climate change [11].

For many decades, demonstrating a direct influence between solar variability and climate change on atmospheric dynamics and in their physical mechanisms in the short term data has proved difficult. There is a long history of solar climate studies, many of which have shown correlations with high statistical significance between solar variability and global surface temperature, e.g. [12, 13, 14, 15, 16], in the globally averaged lower-tropospheric temperatures [17] and in the sea-surface temperature [18, 19]. It has been observed that solar variability strongly influenced the upper stratosphere through an indirect dynamic interaction between ultraviolet radiation (UV) and ozone, e.g. [20] and in the cosmic rays and global amount of low clouds [21, 22], and in particular atmospheric water vapor distribution using NCEP/NCAR Reanalysis data sets [23]. Some correlations between Earth Radiation Budget (ERB) and regional precipitation [24] and between annual precipitation and the variation of sunspot numbers (SSN) [25] have also been observed. Many of these studies employed the SSN and total solar irradiance (TSI) to infer the long-term variability of solar activity to correlate with climate variability, and most of these measurements were mainly made at mid-latitude regions. The total solar irradiance variability during a solar cycle is about 0.1% (maximum) and rapid fluctuation is very small to affect the climate [26]. Alternatively, if the solar radiation in this trend was marked small, it is important to know how these small changes may impact the lower atmosphere such as surface temperature as well as water vapor variability. Thus, many efforts have been made to understand, where and how solar activity exerts their influence in the climate system. Such investigation solar modulation on atmospheric processes is conducted through the poles.

The polar regions being in the forefront of some important climate change issues. To date Antarctica (high latitude Southern Hemisphere) with unique situation in terms of weather and climate has been under-exploited relative to the Arctic (high latitude Northern Hemisphere). Due to its pristine, majestic and unique environment, Antarctica has a privileged position for the study of solar modulation on global atmospheric changes. Their higher latitude regions are directly affected by the onset of the solar particles and energy [27]. Antarctica also provides a key to explain the causes and predict the changes in the global climate system, and in particular to understand global climate change processes. With unique weather and natural environment on Earth, it has great influence on the world's climate. Accordingly, Antarctica acts as a global barometer and has become a base for international weather laboratory in collecting meteorological and space data. However, since the remote location of the southern polar region, the atmospheric water vapor has irregular distribution and difficult to accurately modeling with the use of conventional radio sounding technology. For example, precipitation over Antarctica is still not well determined [28]. Traditional moisture sensors are less reliable in a cold environment [29] to measure the water vapor content. In addition to the data accessibility, lack of measurements in temporal resolution cause difficult to accurately model their variations. Therefore, water vapor remains one of the most poorly characterized meteorological parameters. To overcome these limitations, improved knowledge of atmospheric change monitoring through remote sensing water vapor that is now being addressed using the GPS technology.

Our concerns are about the Global Navigation Satellite System (GNSS) particularly Global Positioning System (GPS) measurements utilized as an essential tool for meteorological or climatic applications. The GPS technique will better in the analysis which providing an increased spatial and temporal resolution of atmospheric data and has proven to be effective-cost on the improvement of weather forecasts and offer essential tool in detecting global climate change. Although GPS receivers were originally installed for geophysical and geodetic applications, they have been employed to accurately measure the atmospheric precipitable water vapour (PWV) and the ionospheric total electron content (TEC) at fine temporal and spatial resolution. PWV, as a climate variable, is one of the key components in determining and predicting the global climate system [30]. At present, numerous projects had been under way in which the impacts of remote sensing of water vapor content in the atmosphere on the improvement of weather forecasts with the GPS are intensively studied [31, 32, 33, 34]. The comparisons of GPS PWV with radiosondes and water vapor radiometers have demonstrated GPS-sensed PWV has achieved an accuracy of about 1~2 mm [35, 36, 37, 38]. More recently, there are many excellent research opportunities for solicits input need to develop, refine and validate the quantification of retrieving atmospheric water vapor uses for improved NWP [39]. This study focuses on the extended ground-based GPS methodology for continuous PWV monitoring that can potentially be used as a standard tool in later solar-weather/climate studies. Thus, improved knowledge of atmospheric change monitoring and modeling will be of high relevance over the coming decades.

This work presents the determination of PWV in Antarctica based on GPS sensing, and investigation of solar influence on PWV. Although the potential role of solar variations in modulating actual climate variability is still an open debate for many decades, it is important to investigate these relationships as recently the Sun's activity was closed magnetically [27]. Such the relationship is required for optimal understanding solar-climate relationships as well as the key to advance our understanding of upper-lower atmospheric interactions towards improved performance of prediction of global climate system. In this work, solar related events influence on water vapor during the high heliogeophysical disturbances, for example during a geomagnetic storm are investigated. A new approach for determining and quantifying the solar influence on PWV by indirectly correlating the TEC and PWV variations on a time basis is proposed. TEC as ionospheric variability associated with the geomagnetic activity of the Sun communicated via solar wind and solar flares, which produces the electromagnetic radiation and the coronal mass ejection, finally affecting the Earth's magnetosphere, ionosphere and in the atmospheric dynamics, is employed as a measure of solar activity, while PWV is used as the terrestrial response. Both quantities are used to implement the TEC-PWV method of ionospheric-tropospheric coupling. This chapter also discusses some of possible mechanisms whereby the Sun may produce significant impact on the PWV variability.

Chapter 2

PWV MEASUREMENTS USING THE GPS METEOROLOGY

The current GPS constellation consists of 32 satellites with nominal operational constellation consists of 24 satellites in six orbital with an inclination of 55° [40, 41]. The orbit of each GPS satellite is approximately circular and semi-synchronized and has placed at 20,183 km from the mean surface of the earth with orbital radius of 26,600 km. For this constellation produces global coverage 24 hours a day. Thus, any location on earth will see roughly 8~12 satellites or 1/3 of the satellites at any one given time. Through a 10 min intervals (a reasonable timescale for large ionospheric and tropospheric dynamics), the GPS line-of-sight (LOS) respect to the ground receiver's field-of-view (FOV) only moves 3° in 20 separate measurements, which is capable for temporal change. Based on these orbital characteristics, PWV measurements from ground receivers at rugged environment and extreme weather can easily detect their temporal variation within the troposphere.

The GPS satellites is emitted their signals in the radio range at L-band frequency and has two modulated frequencies of 1575.42 MHz as L_1 and 1227.6 MHz as L_2. For powerful civilian signal used in the near future, the Aeronautical navigation band of L5 frequency (1176.45 MHz) will be added on L_1 and L_2 [42]. Using a dual-frequency receiver, the basic observable of the GPS signals transmission, that are pseudorange (code) and carrier phase can be measured accurately. The code (ρ) and carrier phase (Φ) observables can be expressed after [43], as

$$\rho = A + a + c(B - b) + D + I + (E + e) \quad (1)$$

$$\Phi = A + a + c(B - b) + \lambda N - I + D + (E + e) \quad (2)$$

where A and a are true range and orbit correction, c is the speed of light, B and b are clock corrections for satellite and receiver, D is the tropospheric delay, I is the ionospheric delay, E and e are the electronics time delay with the satellite and receiver and multipath errors. In Eq. (2), λ and N are the ambiguity term, which λ is the carrier wavelength (m/cycles) and N is the integer carrier phase ambiguity between the GPS receiver and the satellite (in cycles).

Since the GPS signal propagates through the Earth's atmosphere, it is affected significantly by the variability of the refractive index of the ionosphere and troposphere. It causes an excess delay of the signal, and the change in the refractive index with height causes the bending of the signal. In that case, the total delay along the slant path is composed of two parts: ionospheric delay (I) and tropospheric delay (D). The ionospheric delay is frequency dependent and it can be nearly corrected by observing both of the frequencies transmitted by the GPS satellites (L_1 and L_2). The remaining delay is the tropospheric delay or neutral delay, and is not frequency dependent. As such, this parameter along the path of the radio signals to a receiver on the ground is sensitive to local surface pressure, temperature and water vapor constituents. By differencing the pseudorange at two different frequencies, the errors caused by clock biases between receiver and satellite or non-dispersive components in tropospheric delay calculation can be isolated. When all system related errors are accounted, the residual error is presumed to come only from the neutral atmosphere. Delays by neutral atmosphere (i.e. tropospheric delay) are composed of hydrostatic and wet delays. The hydrostatic delay is mainly caused by the dry part of the gaseous in the atmosphere and the non-dipole component of water vapor refractivity. The wet delay part is caused solely by the components of dipole moment and dipole orientation of the water refractivity along the paths of the radio signals to all satellites in view and is highly variable with space and time.

2.1. Total Tropospheric Delay Derived from GPS

Figure 1 shows the flow diagram for the determination of total tropospheric delay and PWV from the measured GPS signal and the surface meteorological data [30, 44]. The total delay caused by the neutral atmosphere, known as the zenith tropospheric delay (ZTD) has two effects on the GPS signal, the zenith hydrostatic delay (ZHD) and the zenith wet delay (ZWD), which are typically computed in units of length (meters). On the other hand, the ZHD measurement is used to correct the errors caused by atmospheric delays of GPS transmissions. While GPS derived ZWD is to obtain the precise PWV content in the atmosphere. The parameters $m_{hyd}(\theta)$ and $m_{wet}(\theta)$ shown in the equation for the calculation of ZTD in Fig. 1 are the hydrostatic and wet mapping functions which are used to reduce or map the dependence of the zenith delay to the satellite elevation angle. In this work, for the practical purpose of calculating the ZTD, the atmospheric layer is considered to have azimuthal symmetry.

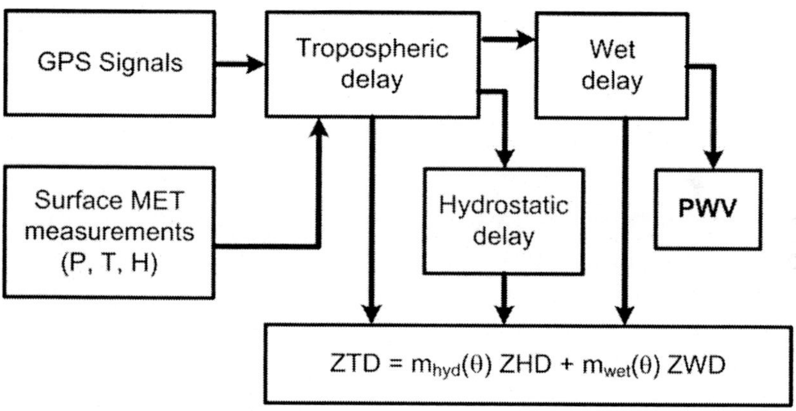

Figure 1. Flow diagram showing the determination of ZTD and PWV [30].

Introducing the lengths of position vectors instead of height to correct the Hopfield model for the determination of refractivity and denoting the Earth's radius by R_E, the corresponding lengths are $r_{hyd} = R_E + h_{hyd}$ and $r = R_E + h_{wet}$ as shown in Fig. 2. R_E is taken as 6378137 meters in this paper. The empirical representation of refractivity to the Modified Hopfield model, N_j as a function of height h above the surface can be written as [30, 44],

$$N_j^{Trop}(r) = N_{j,0}^{Trop}\left(\frac{r_j - r}{r_j - R_E}\right)^4 \quad (3)$$

where for the hydrostatic refractivity component subscript j is replaced by *hyd* and for the wet refractivity component subscript j is replaced by *wet* respectively. In the equation, N_j^{Trop} represent the refractivity above the earth surface and $N_{j,0}^{Trop}$ is the refractivity at the surface of the earth.

In this work, assuming that a non-ideal gas represents the neutral atmosphere layer for the determination of refractivity $N_{j,0}^{Trop}$ in Eq. (3), is corrected by taking the dry (Z_{dry}^{-1}) and wet (Z_{wet}^{-1}) inverse compressibility factors into account. Both the formula for Z_{dry}^{-1} and Z_{wet}^{-1} have been determined empirically by Owens [45] as follows:

$$Z_{dry}^{-1} = 1 + P_{dry}\left[57.97x10^{-8}\left(1 + \frac{0.52}{T_K}\right) - 9.4611x10^{-4}\frac{T}{T_K^2}\right] \quad (4)$$

$$Z_{wet}^{-1} = 1 + 1650\frac{P_{wet}}{T_K^3}\left(1 - 0.01317T + 1.75x10^{-4}T^2 + 1.44x10^{-6}T^3\right) \quad (5)$$

where T and T_K are the surface air temperatures. P_{dry} and P_{wet} are partial pressures of dry air and water vapor respectively. Both P_{dry} and P_{wet} are in mbar, T and T_K are in Celsius and Kelvin respectively. The partial pressure of water vapor can be obtained from the relative humidity (H) as recommended by WMO Technical Note No. 8 [46] as follows:

$$P_{wet} = \frac{H}{100}\exp\left(-37.2465 + 0.213166T_K - 2.56908x10^{-4}T_K^2\right) \quad (6)$$

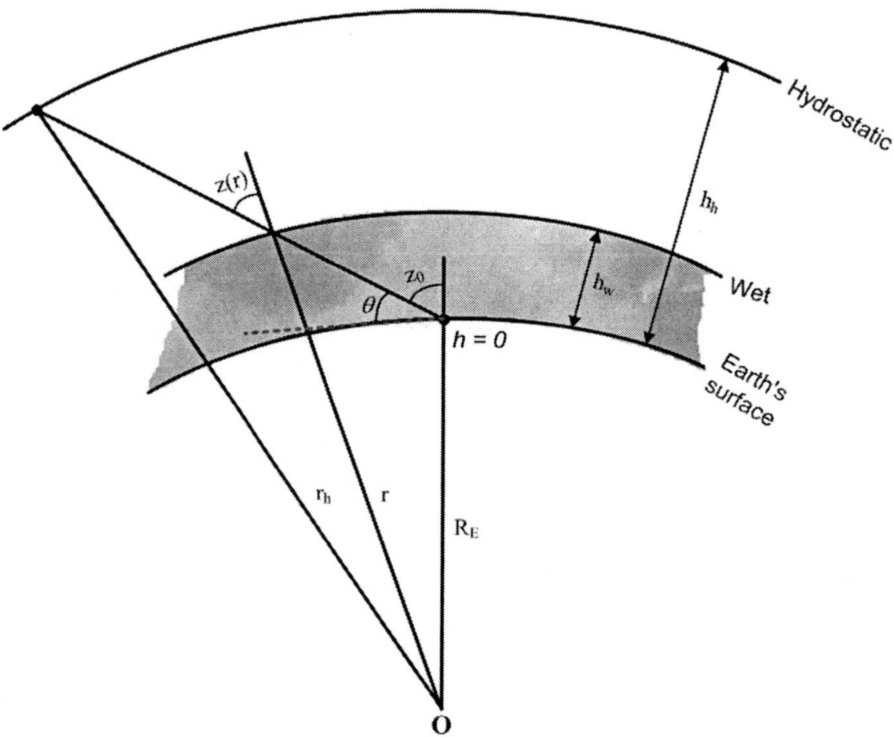

Figure 2. The geometry for the tropospheric path delay based on Modified Hopfield model [30].

By applying the ideal gas equation of state to the dry refractivity component in the Thayer equation [47], $N_{dry} = k_1\left(P_{dry} / T_K\right)Z_{dry}^{-1}$, the dry inverse compressibility factor (Z_{dry}^{-1}) is eliminated. The 'dry' term has been changed to the 'hydrostatic' term, $N_{hyd} = k_1\left(P / T_K\right)$. The total refractivity at the surface of the earth is then given as:

$$N_{j,0}^{Trop} = N_{hyd,0}^{Trop} + N_{wet,0}^{Trop} = \underbrace{k_1 \frac{P}{T_K}}_{hydrostatic} + \underbrace{\left(\underbrace{k_2^{'} \frac{P_{wet}}{T_K} Z_{wet}^{-1}}_{dipole\ moment} + \underbrace{k_3 \frac{P_{wet}}{T_K^2} Z_{wet}^{-1}}_{dipole\ orientation} \right)}_{wet} \quad (7)$$

where P is the measured total atmospheric pressure in mbar. The refraction constants k_1 = (77.6 ± 0.05) K/mbar, $k_2^{'}$ = (22.1 ± 2.2) K/mbar and k_3 = (3.739 ± 0.012) x 10^5 K^2/mbar are computed by Bevis et al. [8]. These constants were chosen because they are more suitable and widely used for deriving delays from both GPS signals and surface meteorological data than the other refractivity constants from other authors [48]. In Eq. (7), the first term is the hydrostatic refractivity in hydrostatic equilibrium and the last term is the wet refractivity component with a new constant $k_2^{'}$ introduced by Davis et al. [49] and is given as

$$k_2^{'} = \left(k_2 - \frac{M_{wet}}{M_{dry}} k_1 \right) \tag{8}$$

In Eq. (8), k_2 is the empirical refractivity constant for dipole moment (70.4±2.2 K/mbar). M_{wet} and M_{dry} are the molar masses of dry air (18.0152 kg/kmol) and water vapor (28.9644 kg/kmol) respectively.

Taking Eq. (3) for the hydrostatic delay and introducing a mapping function $(1/\cos z)$, where zenith angle, $z(r) = 90° - \theta(r)$ is variable and θ is the elevation angle at the observation site as shown in Fig. 2, the ZHD after applying the sine-law can be expressed as

$$ZHD = \frac{10^{-6} N_{hyd,0}^{Trop}}{\left(r_{hyd} - R_E\right)^4} \int_{r=R_E}^{r=r_{hyd}} \frac{r\left(r_{hyd} - r\right)^4}{\sqrt{r^2 - a^2}} dr \tag{9}$$

where the terms in the integral are constant except for r which is variable and $a = R_E \cos\theta$.

Assuming the same model for the wet component, the corresponding formula is given by

$$ZWD = \frac{10^{-6} N_{wet,0}^{Trop}}{\left(r_{wet} - R_E\right)^4} \int_{r=R_E}^{r=r_{wet}} \frac{r\left(r_{wet} - r\right)^4}{\sqrt{r^2 - a^2}} dr \tag{10}$$

The integral in both equations can be solved by a series expansion of the integrand. Adopting the series expansion of Goad and Goodman (as cited in Hofmann-Wellenhof et al. [44]) the solution to the integral r_j is given in Eq. (11).

$$r_j = \left[R_E^2 \left(1 + \frac{h_j}{R_E} \right)^2 - a^2 \right]^{\frac{1}{2}} - \left[R_E^2 - a^2 \right]^{\frac{1}{2}} \tag{11}$$

Solution of the total ZTD (in meters) as a function of θ, P, T, H from Eqs. (9) and (10) can be expressed as Eq. (12) [30, 44],

$$ZTD = ZHD + ZWD$$
$$= 10^{-6} N_{j,0}^{Trop} \begin{bmatrix} 1 + 4a_j \frac{r_j^2}{2} + \left(6a_j^2 + 4b_j\right)\frac{r_j^3}{3} + 4a_j\left(a_j^2 + 3b_j\right)\frac{r_j^4}{4} + \ldots \\ \left(a_j^4 + 12a_j^2 b_j + 6b_j^2\right)\frac{r_j^5}{5} + 4a_j b_j\left(a_j^2 + 3b_j\right)\frac{r_j^6}{6} + \ldots \\ b_j^2\left(6a_j^2 + 4b_j\right)\frac{r_j^7}{7} + 4a_j b_j^3 \frac{r_j^8}{8} + b_j^4 \frac{r_j^9}{9} \end{bmatrix} \tag{12}$$

where,

$$a_j = -\frac{\sin\theta}{h_j}, \quad b_j = -\frac{\cos^2\theta}{2h_j R_E} \tag{13}$$

In Eq. (12), the factor of 10^{-6} was corrected from 10^{-12} in [44] pages of 115 to meet a consistency solution from Eqs. (9) and (10). In Eq. (13), h_j (in meters) represent h_{hyd} and h_{wet} are the effective height for the hydrostatic and wet components, respectively; $h_{hyd} = 40\,136 + 148.72T$ was obtained by fitting global radiosonde data and the tropopause height or wet component (h_{wet}) in this work was set to 11 km. The elevation angle is extracted from the observation of GPS signals.

To achieve high quality of local vertical column of water vapor, the ZHD is calculated using the Saastamoinen (SAAS) model [50]. The SAAS model has become popular among GPS users due to its improved accuracy. It uses the surface pressure measurements and a correction factor $f(\varphi,h)$ to correct the local gravitational acceleration at the center of mass of the atmospheric column. The ZHD SAAS can be expressed (in meters) as follows and is close to unity,

$$ZHD_{SAAS}(P,\varphi,h) = (2.2768 \pm 0.0024)\frac{P}{f(\varphi,h)} \quad (14)$$

$$f(\varphi,h) = (1 - 0.00266\cos(2\varphi) - 0.00028h) \quad (15)$$

where φ is the station latitude (in degrees) and h is the height of the site above the ellipsoid (in km), both are obtained from GPS observation data.

2.2. The Niell Mapping Function

A mapping function is defined as the ratio of the electrical path length (also referred to as the delay) through the atmosphere at geometric elevation, to the electrical path length in the zenith direction [51]. It is developed due to the tropospheric delay is the shortest in the zenith direction and becomes larger with increasing zenith angle. For GPS measurement of zenith PWV, the signal delay in each direction to each GPS satellite is not generally estimated individually. Instead, the individual delays are mapped from each individual satellite direction to a single zenith delay. This mapping method assumes that the delay is independent of azimuth. This assumption could never be made because of the significant increase in delay that is seen when the signal travels through much more of the atmosphere at lower elevations. Mapping functions account the delay for individual satellite view and map them to the zenith direction.

Foelsche and Kirchengast [52] introduce a simple "geometric" mapping function (Fig. 3), where only the free parameter is an "effective height" of the atmosphere, corresponding to about the first two scale height above the surface. The simplest mapping function is the cosecant of the elevation angle that assumes both the curvature of the earth and the curvature of the path of the GPS signal propagating through the atmosphere can be approximated as plane surfaces. This is a reasonably accurate approximation only for high elevation angles with a small degree of bending as given by

$$\frac{ds}{dr} = \frac{1}{\cos z \dfrac{S_{flat}}{S_{atm}}} = \frac{1}{\cos z} \equiv \sec z \quad \rightarrow \; m(\theta) = \frac{1}{\sin\theta} \qquad (16)$$

where dr is the difference in radius (distance to the center of the earth) of the two layers, ds is the distance difference, z is the zenith angle at an arbitrary layer, S_{flat} would be the ray-path within a flat (plane-parallel) atmosphere with assumes $S_{atm} \cong S_{flat}$ (the earth flat and the refractivity is constant).

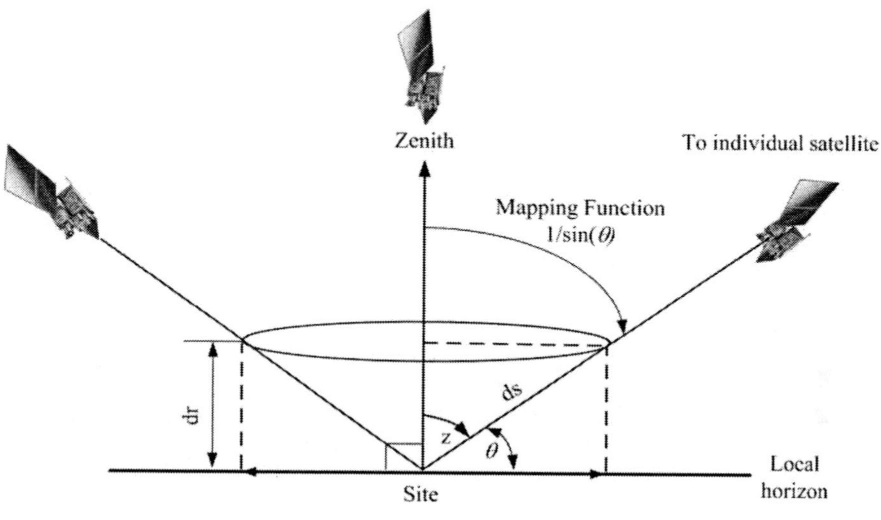

Figure 3. Propagation of GPS signals approximated as a planar surface [48].

Because of the curvature of the atmosphere, this zenith angle changes along the ray-path. Simple mapping function in Eq. (16) is limited for use above ~60 degrees elevation. As the ratio of the thickness of the atmosphere to the radius of the earth decreases, the atmosphere appears more planar. This thickness varies with latitude and season. Thus a possible proxy for the mapping function is some quantity that is a measure of the thickness of the atmosphere. Niell [53] has developed hydrostatic and wet mapping functions with new forms and combination. It is use to reduce errors in geodetic estimation to provide a better fit and give better accuracy over the latitude range 43° N to 75° N for observations down to 3 degrees elevation. The form adopted for Niell mapping function is the continued fraction of Marini [54] with three *a*, *b* and *c* constants. In addition to a latitude and seasonal dependence due to varying solar radiation, the hydrostatic mapping function should also be dependent on the height above the geoid of the point of observation because the ratio of the atmosphere "thickness" to the radius of curvature decreases with height. This does not apply to the wet mapping function since the water vapor is not in hydrostatic equilibrium and the height distribution of the water vapor is not expected to be predictable from the station height. The Niell mapping function for the hydrostatic (replace *j* by *hyd*) and wet (replace *j* by *wet*) components can be written as:

$$m_j(\theta) = \underbrace{m_h(\theta) + \Delta m(\theta)}_{hydrostatic} + m_{wet}(\theta) \tag{17}$$

For the hydrostatic mapping function, the heights above geoid are adjusted. The sensitivity of the hydrostatic mapping function to height above mean sea level (MSL) was determined by beginning the ray-trace with nine different elevation angles between 3 and 90 degrees to give both the hydrostatic and wet path delays of each of the nine standard profiles with the values of pressure, temperature and relative humidity at 1 km and 2 km altitude. The hydrostatic mapping function in Eq. (17) has normalized to yield a value of unity at the zenith and with a height correction, $\Delta m(\theta)$, which can be rewritten as

$$m_{hyd}(\theta) = \frac{1 + \left(a/(1 + b/(1 + c))\right)}{\sin\theta + (a/\sin\theta + (b/\sin\theta + c))} + \left[\frac{1}{\sin\theta} - f\left(\theta, a_{ht}, b_{ht}, c_{ht}\right)\right] h \tag{18}$$

where *a*, *b* and *c* are coefficients to be determined[1], *h* is the height of the site above geoid in meters. Here, $f\left(\theta, a_{ht}, b_{ht}, c_{ht}\right)$ represents the three-term continued fraction in the Marini mapping function,

---

[1] For low precision, the coefficients can be set to a = b = c = 0 which yields the cosecant model as shown in Eq. (16).

$$f(\theta,a_{ht},b_{ht},c_{ht})=\frac{1+\left(a_{ht}/(1+b_{ht}/(1+c_{ht}))\right)}{\sin\theta+\left(\sin\theta+a_{ht}/(\sin\theta+b_{ht}/(\sin\theta+b_{ht}))\right)} \quad (19)$$

The coefficients $a_{ht} = 2.53 \times 10^{-5}$, $b_{ht} = 5.49 \times 10^{-3}$, and $c_{ht} = 1.14 \times 10^{-3}$ were determined by a least-squares fit to the height corrections at the nine elevation angles [53].

For Niell wet mapping function can be written as:

$$m_{wet}(\theta)=\frac{1+\left(a/(1+b/(1+c))\right)}{\sin\theta+(a/\sin\theta+(b/\sin\theta+c))} \quad (20)$$

The coefficients a, b and c in Eqs. (18) and (20) for hydrostatic and wet respectively, are interpolated from latitude of GPS site and takes also seasonal variations into account. It were derived from temperature and relative humidity profiles of the U.S. Standard Atmosphere which is dependent on the latitude at North regions 15° (tropical), 30° (subtropical), 45° (midlatitude), 60 and 75° (subarctic) for the months of January (Winter) and July (Summer). Niell assumes that the Southern and Northern hemispheres are anti-symmetric in time, *i.e.,* that the seasonal behavior is the same. In addition, he assumes the equatorial region is described by the 15° N latitude profile while the polar region are described by the 75° N latitude profile. Tables 1 and 2 show the dependency of coefficients a, b and c on temporal and spatial conditions for the hydrostatic and wet mapping functions respectively.

For the hydrostatic component, these coefficients are determined based on height, latitude and DoY (day of year). However, for the wet mapping function, they depend only on the latitude. The coefficients for hydrostatic mapping function can be interpolated based on the parameter values extracted from Table 1 by the following interpolation rule.

For latitude range $15° \leq |\varphi| \leq 75°$,

$$F(\varphi,t)=F_{avg}(\varphi_i)+\left[F_{avg}(\varphi_{i+1})-F_{avg}(\varphi_i)\right]\frac{\varphi-\varphi_i}{\varphi_{i+1}-\varphi_i}+\ldots$$

$$\left\{F_{amp}(\varphi_i)+[F_{amp}(\varphi_{i+1})-F_{amp}(\varphi_i)]\frac{\varphi-\varphi_i}{\varphi_{i+1}-\varphi_i}\right\}\cos\left(2\pi\frac{DoY-T_0}{365.25}\right) \quad (21)$$

For latitude $|\varphi| \leq 15°$,

$$F(\varphi,t) = F_{avg}(15°) + F_{amp}(15°)\ \cos\left(2\pi\frac{DoY - T_0}{365.25}\right) \tag{22}$$

For latitude |φ| ≥ 75°,

$$F(\lambda,t) = F_{avg}(75°) + F_{amp}(75°)\ \cos\left(2\pi\frac{DoY - T_0}{365.25}\right) \tag{23}$$

where φ is the user's latitude and the subscripts refer to the nearest tabular latitude, F is the mapping function calculated coefficients a, b and c, separated into average values and amplitudes. T_0 is the day of year for "maximum winter" which set to 28 for Northern Hemisphere and 211 for the Southern Hemisphere. The average and amplitude values of the hydrostatic mapping function coefficients are listed in Table 1.

In case of the wet mapping function, the interpolation rule is also based on the following equation but the average values of their coefficients are listed in the Table 2.

For latitude range 15° ≤ |φ| ≤ 75°,

$$F(\varphi,t) = F_{avg}(\varphi_i) + \left[F_{avg}(\varphi_{i+1}) - F_{avg}(\varphi_i)\right]\ \frac{\varphi - \varphi_i}{\varphi_{i+1} - \varphi_i} \tag{24}$$

For latitude |φ| ≤ 15°,

$$F(\varphi,t) = F_{avg}(15°) \tag{25}$$

For latitude |φ| ≥ 75°,

$$F(\varphi,t) = F_{avg}(75°) \tag{26}$$

Table 1. Coefficients of the hydrostatic mapping function

Coefficient	$\varphi = 15°$	$\varphi = 30°$	$\varphi = 45°$	$\varphi = 60°$	$\varphi = 75°$
a_{avg}	$1.2769934.10^{-3}$	$1.2683230.10^{-3}$	$1.2465397.10^{-3}$	$1.2196049.10^{-3}$	$1.2045996.10^{-3}$
b_{avg}	$2.9153695.10^{-3}$	$2.9152299.10^{-3}$	$2.9288445.10^{-3}$	$2.9022565.10^{-3}$	$2.9024912.10^{-3}$
c_{avg}	$62.610505.10^{-3}$	$62.837393.10^{-3}$	$63.721774.10^{-3}$	$63.824265.10^{-3}$	$64.258455.10^{-3}$
a_{amp}	0.0	$1.2709626.10^{-5}$	$2.6523662.10^{-5}$	$3.4000452.10^{-5}$	$4.1202191.10^{-5}$
b_{amp}	0.0	$2.1414979.10^{-5}$	$3.0160779.10^{-5}$	$7.2562722.10^{-5}$	$11.723375.10^{-5}$
c_{amp}	0.0	$9.0128400.10^{-5}$	$4.3497037.10^{-5}$	$84.795348.10^{-5}$	$170.37206.10^{-5}$

Table 2. Coefficients of the wet mapping function

Coefficient	$\varphi = 15°$	$\varphi = 30°$	$\varphi = 45°$	$\varphi = 60°$	$\varphi = 75°$
a_{avg}	$5.8021879.10^{-4}$	$5.6794847.10^{-4}$	$5.8118019.10^{-4}$	$5.9727542.10^{-4}$	$6.1641693.10^{-4}$
b_{avg}	$1.4275268.10^{-3}$	$1.5138625.10^{-3}$	$1.4572752.10^{-3}$	$1.5007428.10^{-3}$	$1.7599082.10^{-3}$
c_{avg}	$4.3472961.10^{-2}$	$4.6729510.10^{-2}$	$4.3908931.10^{-2}$	$4.4626982.10^{-2}$	$5.4736039.10^{-2}$

2.3. GPS Derived Precipitable Water Vapor

PWV is the total quantity of water vapor overlying at a point on the Earth's surface and is expressed as the height of an equivalent column of liquid water. Similarly, Bevis et al. [17] quantified atmospheric water vapor above Earth's surface as the vertical integrated mass of water vapor per area. To calculate the PWV, ZWD can be estimate from the numerical weather model defined as [48]:

$$ZWD = 10^{-6} \int_{h_0}^{\infty} N_{wet}(h)\,dh \qquad (27)$$

In which N_{wet} (see in second term of Eq. 7):

$$N_{wet} = \left(k_2 \frac{P_{wet}}{T_K} + k_3 \frac{P_{wet}}{T_K^2} \right) Z_{wet}^{-1} \qquad (28)$$

By using the gas ideal law for P_{wet} and for non-ideal gases by inserting the wet compressibility factor, the integral for ZWD computation can be written as

$$ZWD = 10^{-6} \left(k_2' + \frac{k_3}{T_m} \right) R_v \int_{h_0}^{\infty} \rho_{wet} Z_{wet}^{-1}\,dh \qquad (29)$$

The right-hand of the integrand in Eq. (29) is known as the Integrated Water Vapor (IWV). For meteorological applications, the ZWD is sometimes expressed in terms of IWV [17].

$$IWV = \int_{h_0}^{\infty} \rho_{wet} Z_{wet}^{-1}\,dh \qquad (30)$$

where ρ_{wet} is the water vapor density along the vertical direction and is defined as the mass of water vapor in a unit volume of air. It is a measure of the actual water vapor content of the air and generally expressed in grams per cubic meter or kg/m^3.

IWV and PWV are related by[1]

$$PWV = \frac{IWV}{\rho_{lw}} \tag{31}$$

where IWV is the integrated water vapor (kg/m^2) and ρ_{lw} is the density of the liquid water (1000 kg/m^3).

The ZWD, however, cannot be sufficiently modeled by surface meteorological measurements due to the irregular distribution of water vapor in the atmosphere. The GPS calculated ZWD (in meters) is obtained by [30]

$$ZWD_{GPS} = ZTD_{GPS} - ZHD_{SAAS} \tag{32}$$

where ZTD_{GPS} is the total ZTD in zenith direction. To obtain the ZTD_{GPS}, in this work, the hydrostatic Niell mapping function is employed. This hydrostatic mapping function is sensitive and suitable for Antarctic regions. Now, the total tropospheric delay in a vertical column (ZTD_{GPS}) can be calculated as [48]

$$ZTD_{GPS} = \frac{ZTD}{m_{hyd}(\theta)} \tag{33}$$

where ZTD is calculated using the Modified Hopfield model with an improvement from Eqs. (12) and (13).

Finally, total PWV (in mm or equivalent depth of liquid water in kg/m^2) is derived from ZWD signals and a conversion factor that proportional to the weighted surface temperature as given after Bevis et al. [8]:

$$PWV = \alpha \; ZWD_{GPS} \tag{34}$$

[1] In many scientific papers, water vapour content is expressed as IWV or Integrated PWV (IPWV), PWV, or only precipitable water (PW). However, in practice, they are related as 1 kg/m2 of IWV corresponds to 1 mm of PWV.

$$\alpha = \left[\rho_{lw} R_v \left(\frac{k_3}{70.2 + 0.72 T_K} + k_2^{'} \right) \right]^{-1} 10^6 = \left(0.102 + \frac{1725.6}{70.2 + 0.72 T_K} \right)^{-1} \quad (35)$$

where R_v is the specific gas constant for water vapor (461.5184 J/mol K). The dimensionless α parameter is a conversion factor that varies with the local climate such as location, elevation, season and weather. From Eq. (34), it is defined that the PWV total is vertically integrated water vapor in a column from a receiver position to the top of the atmosphere.

Chapter 3

Instrumentation, Data Processing and Validation

3.1. Scott Base Antarctic Instrumentation System

Location with optimal conditions should be used so as to have clear sky view, thus allowing reception of low elevation signals. Monumentation of the station should also be stable and highly durable. For the required accuracies to be achieved in the positioning, two main factors have to be taken into consideration in site selection: (i) the position of the station should be determined at the highest accuracy possible by eliminating or reducing all the errors affecting the station position and (ii) the station should be closely spaced since in relative/differential positioning, the same conditions are assumed between receivers at either ends of the baseline. Considering the two above factors, the GPS at Scott Base, Antarctica (SBA) was installed at geographic: 77.85°S latitude, 166.76°E longitude and 28.2 m altitude (Corrected Geomagnetic: 79.94°S, 327.60°E) and the logging system was housed in Hatherton Geosciences Lab. The station was managed by the New Zealand Antarctic Institute, Antarctica New Zealand (ANZ). Figure 4 shows the environment conditions and location of Scott Base, Antarctica (SBA). As shown in the figure, SBA is located at Pram Point, or near the tip of Hut Point Peninsula on Ross Island in the Ross Sea, 1,353 km from the South Pole with Mt. Erebus behind it and at around 3 km from the US base, McMurdo station (MCM) at Discovery Point. Meteorological conditions between SBA and MCM are relatively close to each other, although their topography is quite different. In both stations, annual temperature is about –20 °C with snowfall occurs in all year which maximum during summer.

Figure 4. Environment features and location of Scott Base station in Ross Island, Antarctica. The pictures was taken on 2 February 2008 during the summer 2007/2008 voyage. The American McMurdo base is located at the behind of Observation Hill.

The measurements system at SBA employed for this work consists of a GPS receiving system and a ground-based meteorological system. The GPS was installed in November 2002 under the Malaysian Antarctic Research Programme (MARP). The ground meteorological system was managed by the National Institute of Water and Atmospheric Research Ltd, New Zealand (NIWA) and ANZ. The GPS receiving system is shown in Fig. 5, it consisting of a Trimble TS5700 24-channel, high precision, dual-frequency GPS receiver with a Trimble Zephyr Geodetic antenna and a notebook computer for data logging. The surface meteorological system at SBA consists of a Vaisala HMP45D probe for relative humidity measurement (*H*) in percent and air temperature (*T*) in degrees Celsius, a Vaisala PTB 100A Analog Barometer for surface pressure measurement (*P*) in hPa, a Munro cup Mk11 cup anemometer for the determination of wind speed (*Ws*) in m/s and a W200P potentiometer wind vane for the determination of wind direction (*Wd*) in degrees. The relative humidity and air temperature probe is located inside a standard large wooden Stevenson's screen. The wind sensors are at the top of a 10 m mast.

The pressure sensor is located in one of the SBA laboratories alongside the data logger. The sensors are all connected to a Campbell Scientific CR10X data logger. The sample rate is 3 seconds. Mean data are logged at 10-minute intervals.

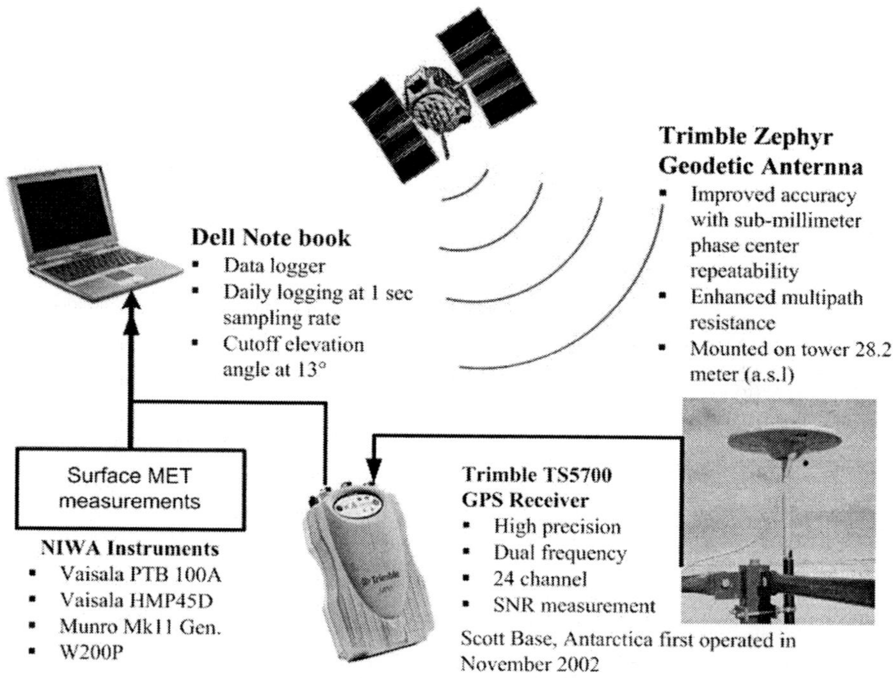

Figure 5. Ground-based GPS PWV measurement system [30].

3.2. Data Set, Processing and Validation Results

To provide information about the performance and limitations of GPS sensed PWV, GPS data at SBA was collected continuously using a Trimble GPS receiver and a Zephyr Geodetic antenna. The surface meteorological data was supported by the NIWA and was maintained by ANZ. At SBA, the GPS receiver was set to track GPS signals with a one-second sampling period and the cut-off elevation angle was set to 13° to eliminate possible multipath effects on GPS data. GPS signals were converted into Receiver Independent Exchange (RINEX) format using the Translate/Edit/Quality Check (TEQC) routine developed by UNAVCO [55]. A 30s sampling period was then used in order to reduce the processing time. A Matlab program suite, namely the tropospheric water vapor program (*TroWav*) was developed by author to process and analysis the above parameters from the data collected. The algorithms of the *TroWav* include satellite elevation angle, ZTD, ZHD, ZWD and mapping function calculations to calculate the PWV. The GPS PWV product at SBA for this analysis has been calculated at 10-minute intervals.

To compare the characteristics of GPS Antarctic water vapor in the polar cap, four sets of GPS and meteorological data in the Antarctic continent were also processed. GPS data at these sites were obtained from the GARNER archives at the SOPAC homepage [56]. All the RINEX observation files in those archives use the Hatanaka format and were processed with a 30s intervals. While the surface meteorological data apart from SBA was obtained from the Australian Antarctic Data Center (AADC) for Casey (CAS1) and Mawson (MAW1), and from the Antarctic meteorology database supported by the British Antarctic Survey (BAS) for McMurdo (MCM4) and Syowa (SYOG) stations. Figure 6 shows the permanent GPS stations in Antarctica. The GPS position and instrument setup together with the manned meteorological sites are listed in Table 3. Based on the availability meteorological data apart from SBA, the PWV data at CAS1 and MAW1 and at MCM4 and SYOG are processed at 1h and 3h averages, respectively.

Using a similar method as at SBA, four sets of GPS and surface meteorological data from other permanent GPS stations were processed to calculate the PWV. The data processed are from the beginning of January 2003 until the end of December 2005. Accuracy of coordinates for the GPS stations is necessary in order to determine the ZHD exactly. In this case, the assessment of the accuracy coordinates used the International Terrestrial Reference Frame, ITRF2002 coordinates [57]. To cancel the residual tropospheric delay, a single differencing technique with baseline length below 10 km was implemented in the pre-processing for precise ZTD estimation.

Before processing the PWV, the total ZTD was validated with ZTD product, namely ZTD reference (ZTDref) [11]. The ZTDref was obtained from the IGS (International GPS service) combined solution provided by the GFZ Potsdam (Geo-ForshungsZentrum, Germany). It represents the weighted mean of the ZTD estimates contributed by 2–6 IGS analysis centers [59]. The ZTDref product was available in 2-hour intervals. Figure 7 shows the comparison between the both ZTD methods and their relationship for the period of October 2003 [58]. Analysis result of Fig. 7 show that a positive bias of ZTD was about 0.48 mm with standard deviation errors to be 0.15 to 0.20 percent. The PWV results were then compared with the PWV obtained from

ZTDref and radiosonde [48]. The PWV difference between GPS and radiosonde is bounded at ± 3 mm, with a RMS value of 1.41 mm. While the PWV difference between GPS and PWV estimated from GFZ had an RMS value of 1.33 mm. From the validation, it clear demonstrated that the estimated PWV between GPS and radiosonde agreed to 5% to each other. the high quality PWV data at SBA had an accuracy of 1.30 mm and a bias of 0.34 mm. In further analysis, the monthly PWV at all stations was processed from daily means to give a clear PWV variation.

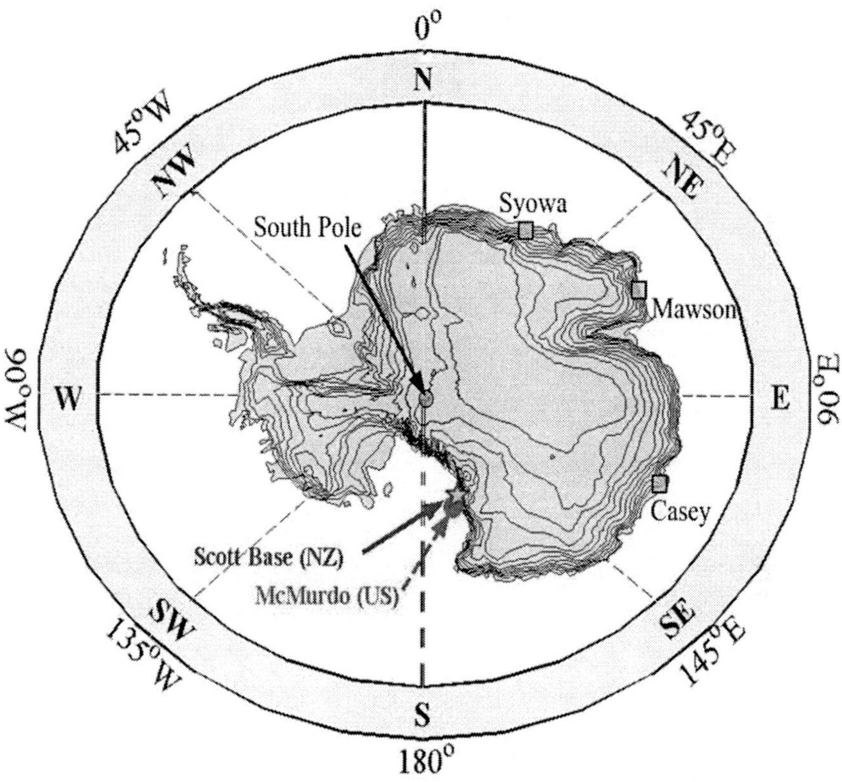

Figure 6. Location of Scott Base station (SBA) and the other fixed GPS sites over Antarctica [27].

Table 3. The geographical coordinates and instrument setup of GPS receivers and manned meteorological (MET) stations in the Antarctic region [27]

Station (country manned)	ID	Latitude	Longitude	Height		Types of GPS	Cutoff elevation angle	Data interval
		(deg)	(deg)	GPS (m)	MET (m)	Receiver and year installed		
Casey (AUS)	CAS1	66.28°S	110.52°E	22.46	50	AOA ICS-4000Z ACT (2003-2004) ASHTECH UZ-12 (2005)	n/a 0°	1h
Mawson (AUS)	MAW1	67.60°S	62.87°E	59.13	16	AOA ICS-4000Z ACT (2003-2004) ASHTECH UZ-12 (2005)	n/a 0°	1h
McMurdo (USA)	MCM4	77.85°S	166.67°E	98.01	24	A00 SNR-12 ACT (2003-2005)	4°	3h
Syowa (JPN)	SYOG	69.00°S	39.58°E	45.00	21	Trimble 4000SSI (2002)	n/a	3h
Scott Base (NZ)	SBA	77.85°S	166.76°E	15.90	16	Trimble TS5700 (2002)	13°	10 min

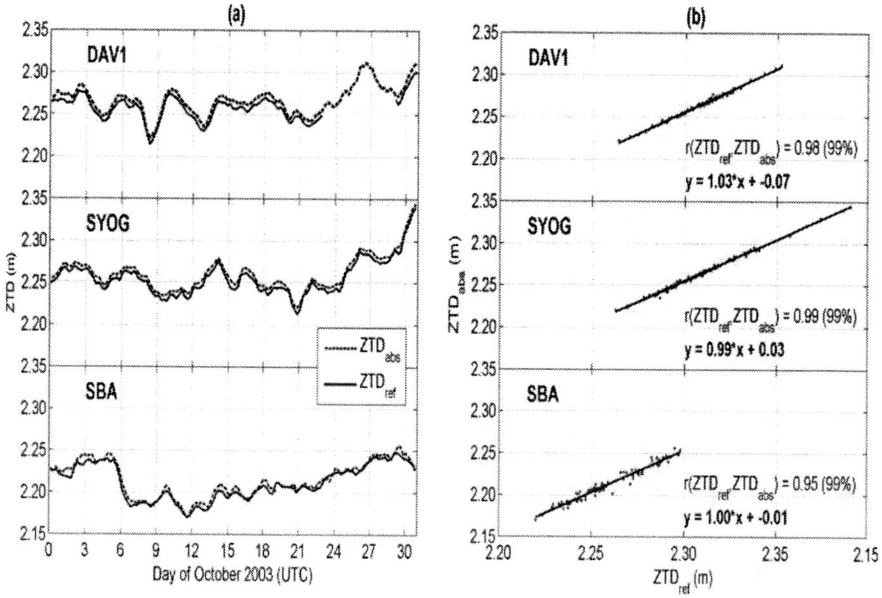

Figure 7. Comparison of ZTD results and their relationship for the period of October 2003 [58].

Chapter 4

CHARACTERISTICS OF SOLAR MODULATED GEOMAGNETIC ACTIVITY DURING THE MAJOR STORMS OF 2003-2005

The 2003-2005 solar extreme events demonstrated the violent and unexpected nature of solar eruptions and provide fundamental insights into the behavior of the Sun and its influence on the space environment of the earth. Four instances of significant solar and severe geomagnetic activities with a minimum Disturbance Storm Time (*Dst*) < -200 nT during the period of 2003-2005 were chosen to study the relationship between solar related events through TEC and atmospheric water vapor. The extreme events are occurred on 29-31 October 2003 (so-called "The October 2003 Storm", cover the period from 27 October to 2 November 2003), on 20-21 November 2003 ("The November 2003 Storm", 17–23 Nov 2003), on 7-10 November 2004 ("The November 2004 Storm", 5–12 Nov 2004) and the last one is on 15-16 May 2005 ("The May 2005 Storm", 13–18 May 2005). Two days of data before and after the storms are analyzed together to give a clear signature of the storm. Table 4 summarizes the solar terrestrial data and flare activities before and after the storms, whereby the locations of flares are identified using H_{α} flares based on NOAA SEC reports. The analyses are presented as follows:

Table 4. Daily solar terrestrial events between 2003 and 2005 during severe geomagnetic storm [30]

Day of events	Solar-magnetic activity				Storm		IMF	Flare activity		
	SSN	$F_{10.7}$	*Ap*	*Kp*[a]	*Dst*[a]	*Time*[b]	*Bz*[c]	*Time*[d]	*Class/Imp*	*Location*
27/10/03	238	257	15	4	-52	05:00, 06:00	-8.24	12:43	M6.7/SF	S17E25
28/10/03	230	274	20	4	-32	10:00, 11:00	-10.36	11:10	X17.2/4B	S16E08
29/10/03	330	279	189	9	-350	24:00	-24.49	20:49	X10.0/2B	S15W02
30/10/03	293	271	162	9	-353, -383	01:00, 23:00	-29.18	02:07	M1.6/1F	N06W22
31/10/03	266	249	93	8	-307	01:00	-5.15	06:18	M1.1/SF	N08W28
01/11/03	277	210	21	5	-69	01:00, 02:00	-3.73	08:22	C4.4/SF	S13W41
02/11/03	174	190	18	4	-41	01:00	-5.20	17:25	X8.3/2B	S14W56
17/11/03	72	121	34	5	-48	07:00	-3.20	09:05	M4.2/1N	S01E33
18/11/03	90	144	20	4	-31	15:00	-2.74	21:52	C2.2/SF	N10E90
19/11/03	114	155	14	4	-27	06:00	-1.55	04:01	M1.7/1N	N01E06
20/11/03	118	175	117	9	-422	21:00, 22:00	-45.60	07:47	M9.6/2B	N01W08
21/11/03	131	177	39	7	-309	01:00	-6.00	09:43	C4.3/SF	S23E36
22/11/03	123	176	22	5	-87	23:00	-8.29	18:36	C1.2/SF	N07E26
23/11/03	158	178	21	5	-79	01:00	-7.65	16:04	C1.5/SF	S21E12
05/11/04	83	141	4	3	-15	01:00	-0.42	11:30	M4.0/1F	N08E15
06/11/04	06	29	3	1	-5	20:00 21:00	-2.87	00:34 19:53	M9.3/2N M1.4/-	N10E08 -
07/11/04	94	30	9	7	-128	24:00	-38.18	16:06	X2.0/-	-
08/11/04	93	24	89	9	-373	07:00	-43.83	15:49	M2.3/1N	N08W35
09/11/04	90	27	20	8	-223	22:00	-20.74	17:19	M8.9/2N	N07W51

Day of events	Solar-magnetic activity				Storm		IMF	Flare activity		
	SSN	$F_{10.7}$	*Ap*	*Kp*[a]	*Dst*[a]	*Time*[b]	*Bz*[c]	*Time*[d]	*Class/Imp*	*Location*
10/11/04	50	105	81	9	-289	10:00 11:00	-28.02	02:13	X2.5/3B	N09W49
11/11/04	70	95	3	5	-115	01:00	-3.17	15:54	C1.1/SF	N08W69
12/11/04	52	97	30	5	-109	11:00	-5.18	-	-	-
13/05/05	100	126	27	5	-54	08:00	-4.84	16:57	M8.0/2B	N12E11
14/05/05	91	100	8	3	-39	20:00	-3.88	15:06	C4.0/-	S09W90
15/05/05	69	103	105	9	-263	09:00	-37.0	22:36 24:03	M3.5/- C2.1/-	S16E15 S16E20
16/05/05	70	99	33	6	-116	07:00	-9.78	02:43 09:08	M1.4/- M1.6/-	S16E18 S16E14
17/05/05	45	90	19	4	-97	06:00	-5.46	02:39	M1.8/-	S15W00
18/05/05	46	84	13	4	-70	03:00	-3.63	08:28 14:49	C1.6/- C2.0/-	S16W13 S15W22

[a] Three hour estimated of activity level (*Kp*) from the NOAA/SEC [60] and hourly final *Dst* (nT units) for 2003 and provincial disturbance storm time (*Dst*) for 2004 and 2005 from the World Data Center-C2 for geomagnetism, Kyoto [61] with maximum of *Kp* and minimum of *Dst* values respectively

[b] Occurrence of *Dst* peak (PD) in UTC

[c] Daily minima of the interplanetary magnetic field (IMF) *Bz* component in nT obtained from [62] based on 1-hour level 2 data from the ACE/MAG instrument onboard ACE spacecraft

[d] Peak time occurrence of flares activity in UT from the NOAA/NGDC [63].

4.1. The Major Storm of 29-31 October 2003 ('Halloween Storm')

The period from 29 to 31 October of 2003 (the "Halloween storm") was characterized by extreme solar activity that resulted in a series of intense geomagnetic storms with final *Dst* less than –300 nT. This storm was the greatest storm during the 23rd solar cycle and one of the fastest traveling solar storms in the last two decades [64]. The extreme interplanetary and geomagnetic disturbances in this magnetic storm were related to the eruptive activity of the sun. The event was preceded by high solar X-ray energy, which started to take place one week before the storm event. During that time, several large active regions (AR) were visible on the surface of the sun which produced substantial solar flare activity a week before, with a combination of a large M-class followed by six X-class flares and huge groups of sunspots on the visible solar disk [65]. Figure 8 shows the solar source for the period of intense geomagnetic storm.

The series of flares appeared to be originating from AR486 and AR484, which produced large geomagnetic storms. The maximum solar X-ray energy was recorded on 28 October accompanied by bursts of radio emission and ejection of solar mass [66, 67]. The most intense storm activities occurred on 29 and 30 October, during which two high-speed streams of solar wind originating from the coronal hole produced a large series of magnetic storms [68]. During this time, two large X-flares, X17.2/4B and the X10/2B flare with energies of 1.8 Wm^{-2} and 0.87 Wm^{-2} respectively were recorded from NOAA GOES satellites. These flares lasted for 93 and 24 minutes respectively. Prior to the major flares, other smaller flares were observed as thin ribbons of emission (see, e.g. [69]). In addition to this violent event, there were two occasions of extremely high solar wind speeds (v_{sw}), one on 29 October with v_{sw} > 1,850 km/s and the other on 30 October with v_{sw} > 1,710 km/s, recorded from ACE measurements.

Figure 9 shows the time plot of the variations of *Dst* index, Interplanetary Magnetic Field in z-component (IMF *Bz*), solar and geomagnetic indices for the October 2003 storm including the two days before and after the storm event. Based on *Dst*, the storm has two distinct episodes, preceded by high solar activity. The first storm episode (Ep_1) occurred with the onset of the first sudden storm commencement or SSC_1 from times 06:12 UT to 07:00 UT on 29 October followed by a *Dst* peak value of –180 nT, which was caused by the shock sheath ahead of Ejecta-II [70]. The IMF *Bz* inside the sheath even reached –14 nT. A large-scale magnetic structure with a rotating magnetic field was observed between 11:17 UT on 29 October and 03:50 UT on 30 October.

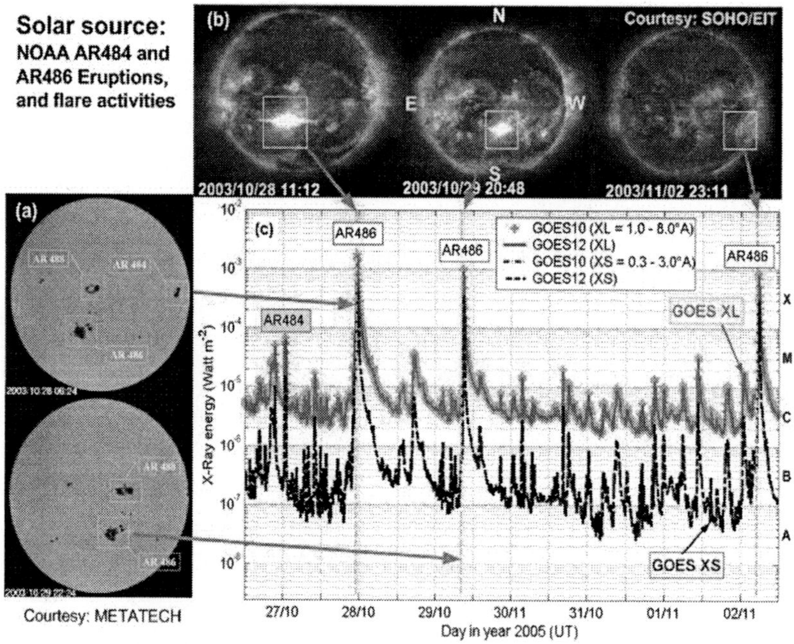

Figure 8. Solar source: (a) Active region 486 and 484 eruptions, (b) enormous of solar flares on 28 and 29 October 2003 from SOHO-EIT at 195 °A and (c) GOES X-ray flare activity from 27 October to 2 November. Note the intensity of X-ray flux in two bands covering XL = 1.0 to 8.0 °A and XS = 0.3 to 3.0 °A as measured by NOAA GOES-10 and GOES-12 satellites respectively [48].

Thereafter, the storm rapidly intensified to a minimum *Dst* (PD_1) of -353 nT at 01:00 UT on 30 October followed by an extended recovery phase and ended at around 17:00 UT. In this episode, a large IMF *Bz* event of about ~-25 nT at 20:00 UT occurred on 29 October. The second storm episode (Ep_2) occurred between 18:00 UT on 30 October beginning at the SSC_2 and ended at about 12:00 UT on 31 October with a minimum *Dst* (PD_2) of -383 nT at 23:00 UT on 30 October. During this event, IMF *Bz* earlier fluctuated to strongly northward almost +23 nT before turning southward to minimum peak value of about ~-30 nT at 21:00 UT ahead of PD_2 which was caused by shock compression of preexisting magnetic cloud material [71], then followed by 3 hour *Kp* index recording maximum value of 9 for four times at 09:00 UT and

21:00 UT on 29 October and at 21:00 UT and 24:00 UT on 30 October. Both negative IMF *Bz* events clearly showed that they were responsible for the magnetic storms on 30-31 October. As shown in the figure, the global planetary index (*Ap*) jumped abruptly from a value of 20 on 28 October to a maximum value of 189 on 29 October and decreased dramatically to a value of 18 on 2 November when the storm subsided. The solar flux index ($F_{10.7}$) exceeded 250 and the sunspot number exceeded 200 a few days before the storm event. Both maintained their high values during the whole storm event indicating that the sun had been in an active condition a few days prior to the storm event.

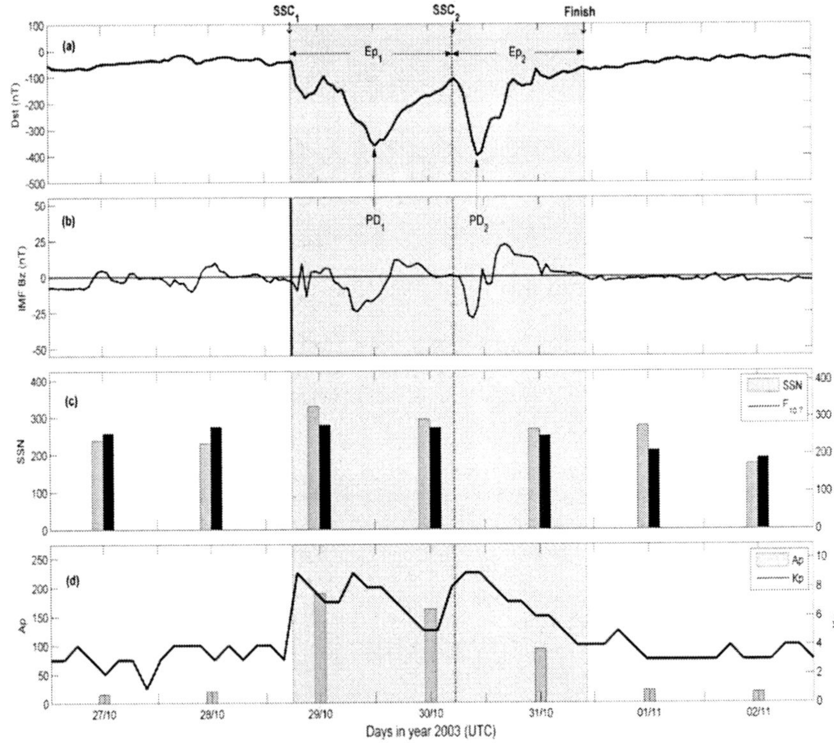

Figure 9. The *Dst* index, IMF *Bz*, solar and geomagnetic indices between 27 October and 2 November 2003. The grey background shows the beginning and ending of the both storms. The vertical solid line in the IMF *Bz* shows the shock due to a magnetic cloud [30].

4.2. The Major Storm of 20-21 November 2003

Another large geomagnetic storm of solar cycle 23 occurred due to coronal mass ejection (CME) from AR501 on 20 November 2003, traced to CME that erupted on 18 November [72]. Figure 10 shows the solar source from AR501 eruption for the period of intense geomagnetic storm. Two M3.2/2N and M5.8/2B moderate flares peaking at 07:52 UT and 08:31 UT were recorded on 18 and on 20 November respectively from the CELIAS/SEM sensor aboard the SOHO spacecraft [73]. These flares were accompanied by a huge CME that looked at first like a bright wide high-speed (1,820 km/s) loop and then developed into a full halo [65]. During this event, the solar wind arrived with a speed of 805 km/s, slowly decreasing to a minimum value of 447 km/s before reaching 733 km/s at the beginning of magnetic clouds as indicated by Gopalswamy et al. [72]. For the period of 20 to 22 November, the measured solar wind speed was not as high as the October magnetic storm. However, the November 2003 magnetic storm developed in the magnetosphere was stronger than the October 2003 storm with a long period of negative IMF *Bz* and with lowest peak *Dst*.

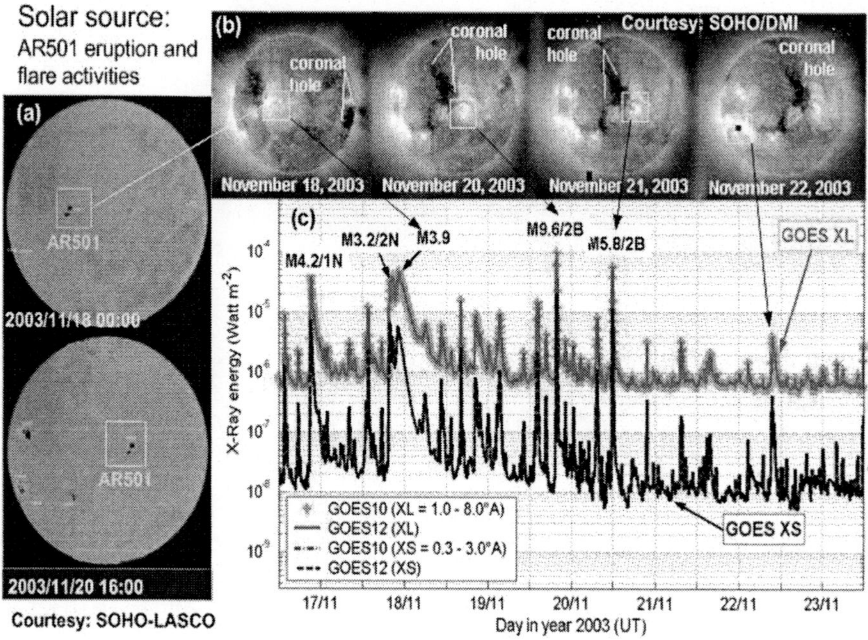

Figure 10. (a) Solar source from AR501 eruption, (b) enormous of flare activity and (c) GOES X-ray flares from 17 to 23 November 2003 [48].

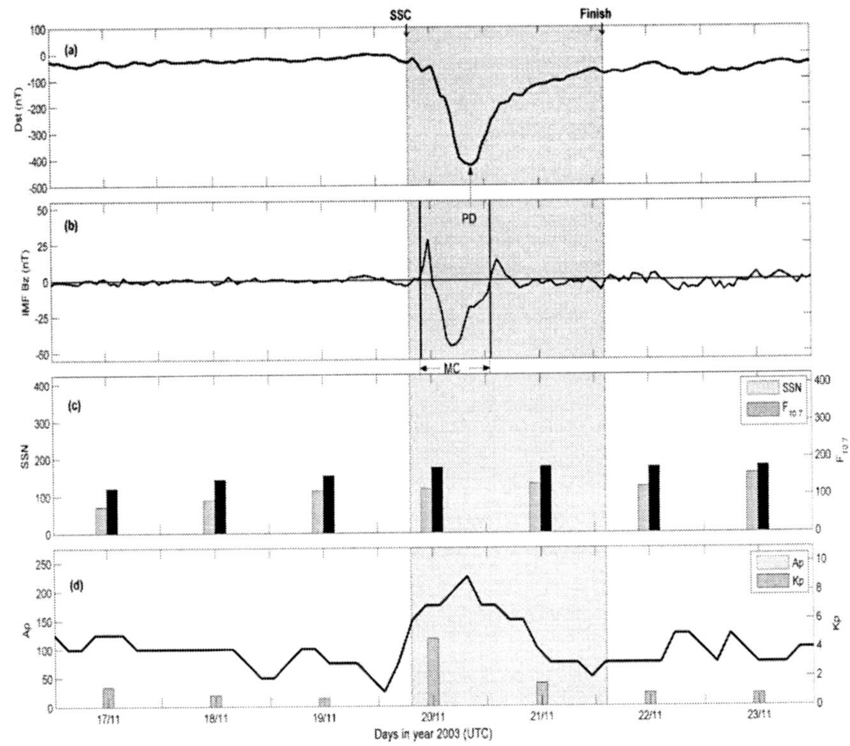

Figure 11. The IMF *Bz*, magnetic and solar indices between 17 and 23 November 2003 [30]. Note that the magnetic cloud (MC) event on 20 November has been identified by Gopalswamy et al. [72].

Figure 11 shows the plot of *Dst* index, IMF *Bz*, solar and geomagnetic indices for the November 2003 storm. The storm based on *Dst* phase started with a rapidly intensifying SSC at around 07:00 UT 20 November associated with solar flares and CMEs with a minimum *Dst* (PD) equal to -422 nT at 21:00 UT. As shown in the figure, measured data from ACE/SWEPAM showed the IMF *Bz* strongly northward turning to +28 nT before the shock front reached the earth at 08:05 UT on 20 November. The CME near the Sun associated with the magnetic cloud contained an unusually strong and long period of IMF *Bz*, which peaked at almost -46 nT. During this event *Kp* was equal to 9 at around 18:00 UT and *Ap* jumped abruptly from a value of 14 on 19 November to a maximum value of 117 on 20 November, decreasing dramatically to a value of 39 on 21 November when the storm subsided. Note that *SSN* and $F_{10.7}$ values increased for the November 2003 storm. *Dst* values less than -300 nT during the period of October-November 2003, clearly indicated that the Sun was most active and violent for the period of cycle 23.

4.3. The November 2004 Storm

The event of November 2004 also indicated a solar source due to geomagnetic activity in which the Sun again put on a fascinating show during the Halloween season with several large flares and CMEs. The largest flares were observed by GOES X-rays: an X2 on 7 November and an X2.5 on 10 November from AR696. These intensities were nearly a factor of 10 times smaller than the historic activity observed during late October 2003. According to [74] three separate CME passages were generated by flare activity from the above referenced sunspot activities. The first two CME passages generated long duration geomagnetic storms, while the last one was minor with little geomagnetic storm activity. The solar wind transferred energy into the Earth's magnetosphere for extended periods of time from 7 to 10 November; speeds ranged from 567 to 809 km/s during the observation period. Figure 12 shows the solar source predominantly from AR696.

Figure 13 shows the plot of *Dst* index, IMF *Bz*, solar and geomagnetic indices for the period of 5-12 November 2004. During the period, there were 2 X-class, 8 M-class and 48 C-class flares are recorded. As shown in the figure, the period of 7-11 November 2004 was very disturbed. The storm based on *Dst* phase has two distinct episodes preceded by a high solar activity, but with a different character as in the October 2003 storm. The first (Ep_1) and second (Ep_2) episodes lasted from 7 to 9 November and from 9 to 11 November respectively. For Ep_1, SSC_1 occurred just after 11:00 UT on 7 November with *Dst* of +42 nT, which then dropped to a minimum value (PD_1) of –373 nT at 07:00 UT on 8 November. During this event, the IMF *Bz* turned southward for a very short duration around 14:00 UT on 7 November and recovered thereafter to become northward by about 15:00 UT and attaining a peak value of about +37 nT. It fluctuated violently between north and south directions for a very short duration and then reduced to a minimum value of almost –45 nT at about 04:00 UT on 8 November.

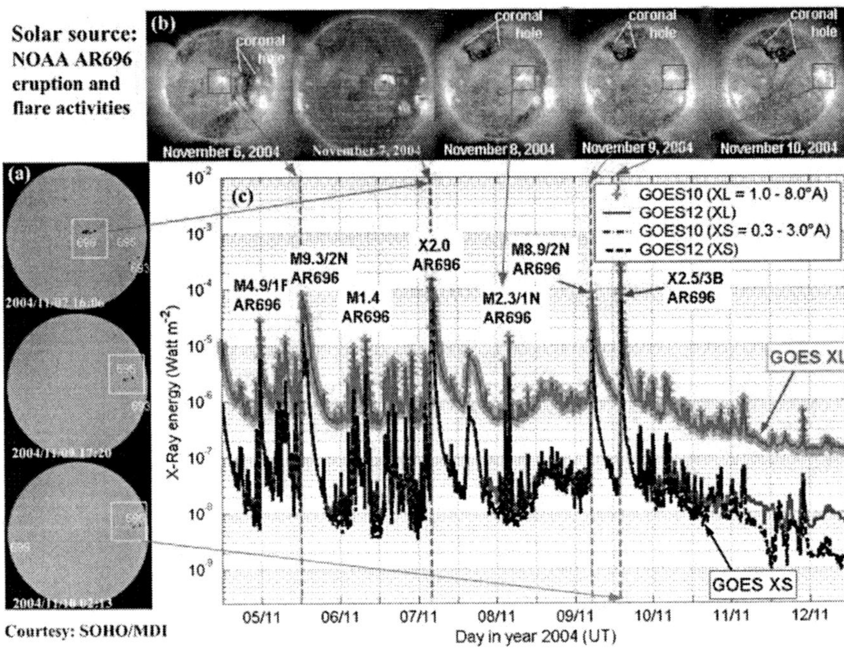

Figure 12. (a) Solar source related AR eruptions, (b) enormous of flare activity from SOHO-MDI and (c) GOES X-ray flares for the period of 5-12 November 2004. The AR in (b) marked by square was indicated a very active flares [48].

As shown in the figure, the *Ap* and *Kp* jumped abruptly to 89 and 9 respectively on 8 November. For Ep_2, SSC_2 occurred just after 12:00 UT, two small substorms with minima *Dst* of –156 nT at 16:00 UT and –223 nT at 22:00 UT on 9 November, followed by a minimum *Dst* (PD_2) equal to –289 nT at 11:00 UT. Thereafter, storm started the recovery phase for a long duration and was almost over around 12:00 UT on 11 November. During this event, the IMF *Bz* became southward for a short time. The IMF *Bz* turned strongly northward to a value of about +40 nT. It remained fluctuating between northward and southward before going southward and reached a minimum value of about –28 nT at 18:00 UT on 10 November. Thereafter the geomagnetic field was at active to extremely severe storm levels. As shown in the figure, *Ap* and *Kp* reached maximum values of 81 and 9 respectively, similarly character to Ep_1 and then decreased when the storm subsided. *SSN* and $F_{10.7}$ values were seen to slowly decrease with the declining phase of solar cycle 23.

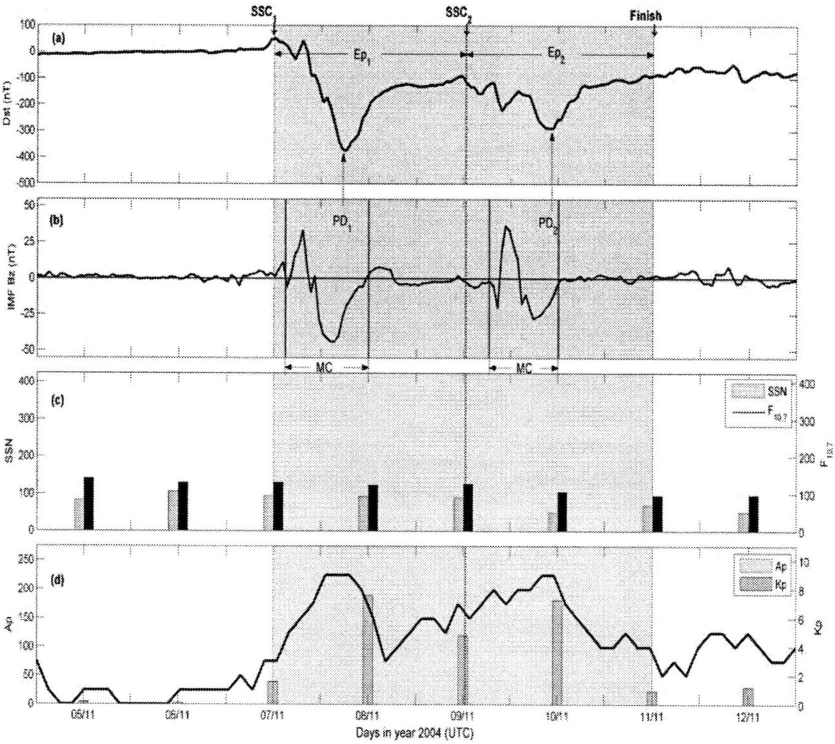

Figure 13. The IMF *Bz*, magnetic and solar indices between 5 and 12 November 2004 [30].

4.4. The Major Storm of 15-16 May 2005

The severe geomagnetic storm in May 2005 is an interesting analysis as the *Dst* peak reached < -200 nT. Figure 14 shows the solar source for the period of intense geomagnetic storm with produces many solar flares for the May 2005 storm. At the top and left panels shows the enormous of flare activities were seen at the SOHO-EIT images from AR759 and 763 eruptions. An unexpected solar burst associated with a proton event with M8.0 class flare between 16:13 and 17:28 UT with peak emission at 16:57 UT on 13 May were recorded from NOAA AR759 with a fast halo CME as shown in LASCO/C2 images. This flare led to a shock 33.5 hours later at 09:15 UT on 15 May [73] and corresponding CME and intense geomagnetic storm arrived at earth as observed at 3:00 UT, preceded by a sudden increase in solar wind speed at 1 AU from 414 to 959 km/s observed by ACE/SWEPAM instrument on ACE. A very strong solar wind shock was also observed at SOHO at 12:18 UT with a sudden increase in solar wind speed from 514 to 971 km/s, also reported by Jan Alvestad [75].

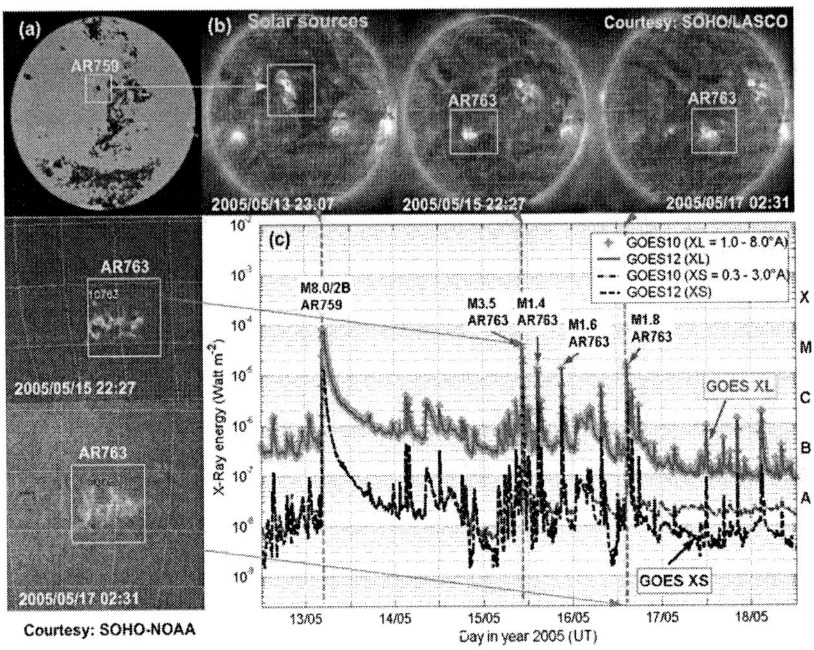

Figure 14. (a) Solar source related AR eruptions, (b) enormous solar burst and (c) GOES X-ray flare activity from 13 to 18 May 2005 [48].

Figure 15 shows the plot of *Dst* index, variation of IMF *Bz* and solar and geomagnetic indices for the period 13-18 May 2005. As shown in the figure, days 15 and 16 of May were the most disturbed by a severe geomagnetic storm. The storm started with a positive *Dst* and then a rapidly intensifying SSC at around 02:00 UT on 15 May associated with IMF and solar wind shock and was almost over at around 16:00 UT on 17 May. For this period, the polarity of IMF *Bz* showed a small fluctuation before turning violently southwards at 05:30 UT. It then turned slowly northwards and was northward after 08:15 UT with value of about +25 nT. The fluctuation in IMF *Bz* between 05:30 UT and 06:10 UT on 15 May prior to the shock caused southward IMF *Bz* due to disturbances in the magnetosphere and it reached the lowest value of about -37 nT driving a superstorm with a minimum *Dst* (PD) of -263 nT at 09:00 UT. An intense substorm was triggered just as IMF *Bz* reached its lowest negative values. Thereafter, the recovery phase started and ended at around 16:00 UT on 17 May. During this event, *Kp* suddenly reached a value of 5 during the SSC and maximized at a value of 9. *Ap* obtained from NOAA jumped abruptly from a value of 8 on 14 May to a maximum value of 105 on 15 May and decreased dramatically when the storm subsided. As shown in the figure, *SSN* and $F_{10.7}$ had similar trends to the November 2004 storm due to the current solar cycle going in descending phase. In summary, two major storms for *Dst* < -200 nT and *Kp* ≥ 7 were recorded during the period. Due to the strong southward IMF *Bz*, the CME on 13 May produced a strong geomagnetic storm (*Dst* = -263 nT) and it was associated with the magnetic cloud structure. The MC arrived at about 06:00 UT and it ended at about 19:12 UT (its boundaries indicated by vertical lines).

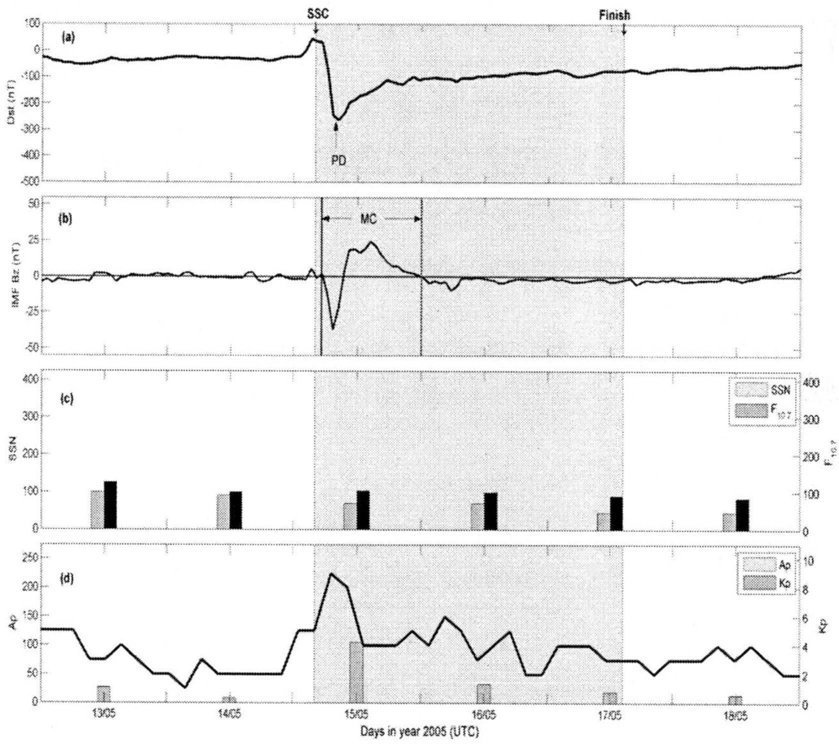

Figure 15. The IMF *Bz*, magnetic and solar indices between 13 and 18 May 2005 [30].

Chapter 5

A New Proposed Method to Correlate Solar Activity and Terrestrial Climate

It is now well established that TEC as ionospheric variability is associated with the geomagnetic activity and solar flares originated from the Sun. Rishbeth [76] found that long-term changes in the ionospheric parameters vary systematically with time of day, season and latitude and are affected by solar and geomagnetic activity. Many modeling techniques have successfully described long-term ionospheric variations, such as that directly related to the sunspot cycle and their results are useful in predicting monthly median values of basic ionospheric characteristics for a given place and time [77]. These indicate that the ionospheric TEC variability is a direct manifestation of the Sun activities. The TEC also seems to have effects on PWV at high latitudes.

Using the argument that solar events have a direct and clear influence on TEC, we are able to relate the influence of solar events on PWV indirectly by correlating ionospheric TEC with PWV measurements. In the proposed method, solar influence on PWV is determined indirectly by correlating ionospheric TEC with PWV measurement on a time-basis. In this assumption, TEC and PWV are represented as a measure of solar activity and terrestrial climate response respectively and for simplicity so-called "GPS TEC-PWV method". TEC-PWV from GPS signals act as a wire to linkage and brings the energy source from the Sun radiation via changes solar activity to the Earth's atmosphere and this source to have potential effects on the meteorological parameters in the tropospheric layer. This argument is also valid because both the ionospheric and the tropospheric layers or the geospace (near-Earth environment) are closely coupled, interact continuously and both layers in some ways affect the GPS signals [30]. Such a method has not been investigated before and we believe this could open a window for better analysis and understanding of the solar-climate relationship. To characterize the solar-climate relationship based on TEC measurement, a strong event is required, as it would minimize the influence of unwanted effects or noise. Such a strong event is a geomagnetic storm. However, extra care is needed in interpreting the data since on the ground the kinetic energy of the wind can have a strong influence on the distribution of PWV as will be demonstrated.

In this study, the GPS TEC was calculated based on Hofmann-Wellenhof et al. [44] and Warnant and Pottiaux [78] methods, which can be found in Abdul Rashid et al. [79].[1] For short-term correlation between PWV and TEC, the percentage deviation of TEC (%ΔTEC) was employed in the analysis rather than the absolute TEC, as this provided the TEC deviation (enhancement or depletion) from the mean value in response to the solar forcing on the ionosphere. The percentage of ΔTEC in this work was adapted from Danilov and Lastovicka [80]. However for consistency we use the mean quiet-day instead of the median quiet-day as given by [30]

$$\Delta TEC\ (\%) = \frac{TEC_{\text{obs}} - TEC_{QD}}{TEC_{QD}} \text{x} 100\% \tag{36}$$

where TEC_{obs} and TEC_{QD} are the storm time and mean quiet-day TECs respectively. Both TEC_{obs} and TEC_{QD} are in TECU (1 TECU = $1\text{x}10^{16}$ electrons m^{-2}).

To quantify PWV in short-term analysis and its correlation with %ΔTEC, the lower threshold value of PWV (PWV_{th}) is set to 5 mm, where PWV level greater than or less than this threshold value indicates the influence of the surface meteorological condition and/or the coupling of the geomagnetic storm (solar activities) to the atmosphere. Note that the PWV_{th} is determined from daily mean PWV for the year of 2003 and 2004 during clear sky and quiet storm conditions at SBA and MCM stations. On the other hand, value of $Kp \geq 3$ has been widely accepted as an indicator for an active geomagnetic storm condition. The %ΔTEC is also a good indicator of a storm activity. Assuming

[1] In this work, the TEC term is used instead of Vertical TEC (VTEC) for simplicity. All the VTEC values used in this work have been corrected using the satellite and receiver bias values obtained from the data center of Bern University, Switzerland.

that the measured data are not free from errors, an error margin of 10% for %ΔTEC and 0.25 mm for PWV were introduced in the data analysis where any deviation from these margins reflects the real signature of the measured parameters and the various methodological such as due to the possibility of incorrect receiver bias. In other word, both %ΔTEC and PWV margin error are defined as adjustment threshold values. In detail, the value of 0.25 mm is obtained from mean difference of PWV between conversion factor α of Eq. (35) calculated using a surface temperature and it was set to ~0.15 [8], while 10% for %ΔTEC is obtained from mean RMS of TEC. Wind speed threshold (W_{th}) $\geq$ 5 m/s and humidity (H) $\geq$ 50% are also used to select the strong PWV and TEC signals to obtain a better insight on the solar influence.

Chapter 6

Investigation of Solar Influence on Terrestrial Climate

The indirect influence of solar related events on the production of PWV was investigated by correlating ionospheric TEC (percentage deviation of TEC) with PWV measurements. Four subsections discuss, firstly, local meteorological conditions at specific location that Scott Base and McMurdo station are described. Secondly, the solar influence on PWV in the short-term during investigation four severe geomagnetic storms. The third compares solar activity and PWV during quiet days based on hourly averages and the fourth presents the relationship between solar activity and PWV based on monthly data. The data analyzed are from January 2003 to December 2005.

6.1. Surface Meteorological Conditions During Geomagnetic Storms

Figure 16 shows the local meteorological conditions at SBA and MCM during the October 2003 storm, the November 2003 storm, the November 2004 storm, and the May 2005 storm periods. As shown in the figure, the event at MCM led SBA by about 3h. The mean values of the measured mean sea level surface pressure at SBA in all storms were higher than MCM by about 0.05%. The measured temperatures at SBA were also higher by about 20% and 23% for the October 2003 storm (Fig. 16a) and the November 2003 storm (Fig. 16b), respectively. However, for the November 2004 storm (Fig. 16c) and the May 2005 storm (Fig. 16d), the mean temperatures at SBA were lower than MCM by about 2.9% and 8.8%, respectively. The measured humidity at both stations for the first two storms dropped before and after the storm event with a mean value of humidity at SBA higher than that of MCM by about 13%, 12.86% for the November 2004 storm (Fig. 16c) and 5% for the May 2005 storm (Fig. 16d). The local wind speed at both stations in Fig. 16a was moderate compared to the maximum speed in the November 2004 and May 2005 storms. In Fig. 16b, the local wind was from the Northeast at both stations with a maximum speed of about 7.36 m/s and 8.20 m/s respectively. While in Fig. 16c, the local wind was from North and East with a maximum speed of 17.13 and 35.0 m/s at SBA and MCM respectively. Note that as shown in the figure, the wind speed at MCM is about twice than at SBA. In Fig. 16d, the wind at both stations was observed to flow from the Northeast with maximum speeds of 21.56 and 35.0 m/s respectively.

Referring to Fig. 16a, the surface pressure prior to the storm event reached a maximum peak of 998.2 mbar and 997.2 mbar for SBA and MCM respectively, while the temperature decreased to -21.8 °C and -19.8 °C and the humidity reached values of 58% and 49% respectively at both stations. This event coincided with the occurrence of a X17.2 class flare. At SBA, local wind with a speed of 9.45 m/s was observed to flow from the Northeast direction with decreasing speed. A similar trend of wind speed was observed at MCM. At the onset of the SSC_1 the wind speed begin to increase again at both stations about the same time and reached a maximum speed of 10.6 m/s for SBA and 5.1 m/s for MCM before decreasing. After the onset of SSC_2, wind speeds began to increase again, at MCM, it started at about one hour after the onset of SSC_2 but at SBA, it occurred 2 hours before the onset of SSC_2. After the storm period, wind speeds at both stations began to decrease with an average value of 1.67 m/s, flowing from the Northeast direction. Prior to the storm event in Fig. 16b, the average maximum wind speed was ~6.30 m/s from the Northeast. Both pressures linearly increased by about 0.44 mbar, the temperature was almost stable with an average of –11°C and the humidity reached a maximum peak of 87% for SBA and 76% for MCM on the afternoon of 18 November. At the SSC, both pressure and temperature linearly increased, while the humidity reached a maximum value of 75% for SBA and 80% for MCM; during this time, the wind speed was 4.49 m/s at SBA and 5.60 m/s at MCM. During the storm period, the average local wind speed measured at SBA was 5.09 m/s, while at MCM it was equal to 2.80 m/s from North-Northeast and it decreased to a minimum value, in contrast with the storm on 21 November. At 14:00 UT on 22 November (marked finish on the graph), the wind speed dipped to a value of 5.48 m/s at SBA and 3.60 m/s at MCM; after

2 hours their speed at both stations increased to a maximum peak of 5.52 m/s and 6.70 m/s respectively.

Referring to Fig. 16c, the surface pressure at SBA and MCM prior to the storm event increased with a gradient of 0.63 mbar in 4 days, while the temperature slowly decreased with an average value of –10.87 °C and the humidity reached a value of 90.4% and 84.0% respectively at both stations. At MCM, the local wind speed with three gradual peaks, the highest being 31.0 m/s, was observed to flow from the North. A similar trend of wind speed was observed at SBA, but quite low, reaching a maximum of 17.13 m/s, from the Northeast. At the onsets of the SSC_1 and SSC_2, the wind speed at both stations began to decrease to 1.24 and 7.00 m/s respectively.

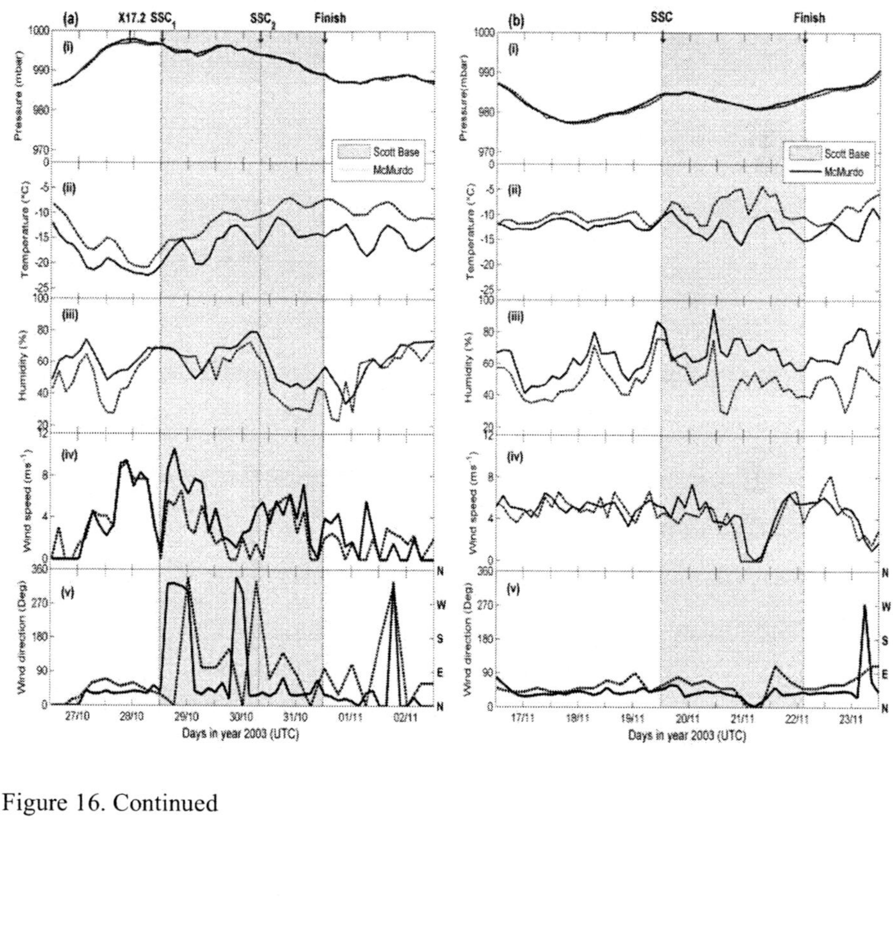

Figure 16. Continued

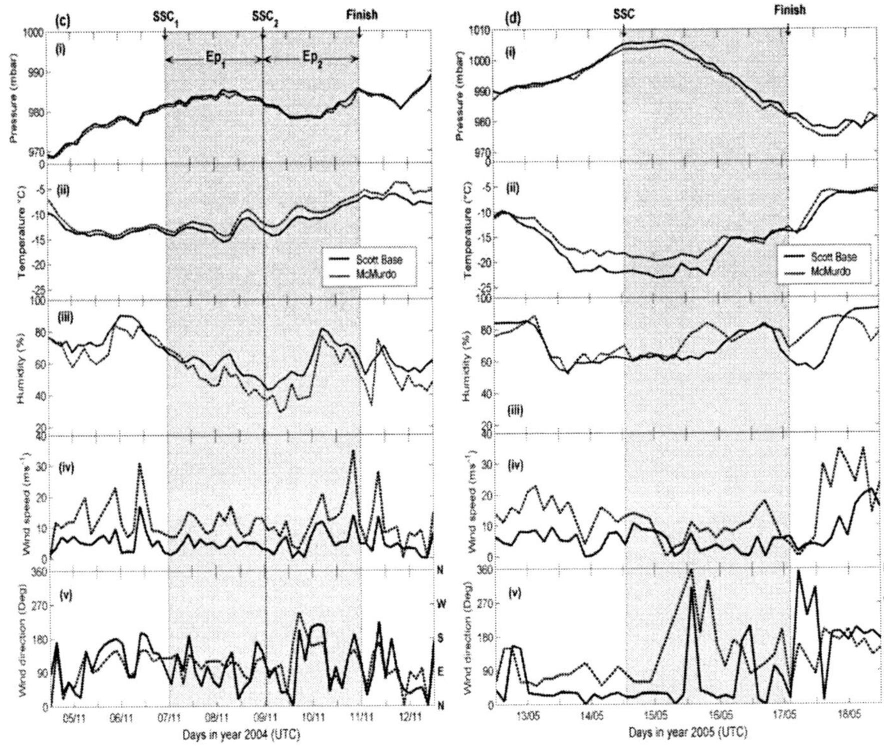

Figure 16. Time series of surface meteorological measurements for (a) the October 2003 storm, (b) the November 2003 storm, (c) the November 2004 storm and (d) the May 2005 storm periods at SBA and MCM stations. Local time for both stations is UT + 12h and the dates marked on the figure represent noon UT [30].

After the storm period, wind speeds at both stations began to increase with maximum values of 13.5 and 28.0 m/s respectively, both from the East to South. For the May 2005 storm in Fig. 9d, the surface pressure prior to the storm event was increased linearly with gradient equal to 0.90 mbar in 3 days on average. The average of temperature and humidity at both stations decreased from –8.11 to –22.60 °C and 87.90 to 49.83% respectively. The local wind at MCM showed high variability with standard deviation of 6.57 m/s and at SBA about 3.33 m/s; both winds were from the Northeast. At the SSC, surface pressure increased to an average value of 1004.3 mbar. Both temperature and humidity were almost stable with temperature readings equal to –26.03 °C and –24.22 °C, while the humidity was 58.87% and 70.0%,

Table 5. The characteristic of surface meteorological conditions during the major storms of 2003-2005 [30]

Station	Average value				Peak value			Remarks
	P^a (mb)	T (°C)	H (%)		P^a (mb)	T (°C)	H (%)	
(a) The October 2003 storm								
SBA	994.24	-15.61	60.89		998.20 mb at 12:40 UT on 28 October	-12.12°C at 11:00 UT on 31 October	79.15% at 18:00 UT on 30 October	Lagging
MCM	994.09	-11.55	54.47		997.20 mb at 10:00 UT on 28 October	-9.82°C at 09:00 UT on 31 October	71.78% at 15:00 UT on 30 October	Leading
Station difference	0.15	-4.06	6.42		1.00 mb (-1.2h)[b]	-2.59°C (-2h)[b]	7.37% (-3h)[b]	SBA lagging MCM
(b) The November 2003 storm								
SBA	983.23	-12.46	68.64		985.30 mb at 12:00 UT on 20 November	-9.12°C at 03:00 UT on 20 November	94.76% at 21:00 UT on 21 November	Leading
MCM	982.87	-8.46	51.20		985.10 mb at 12:00 UT on 20 November	-4.30°C at 18:00 UT on 21 November	75.40% at 00:00 UT on 20 November	Lagging
Station difference	0.36	-4.00	17.44		0.20 mb (0h)	-4.82°C (35h)[b]	19.36% (-38h)[b]	SBA leading MCM
(c) The November 2004 storm								
SBA	982.14	-12.11	60.70		985.83 mb at 15:00 UT on 8 November	-9.72°C at 03:00 UT on 9 November	83.90% at 15:00 UT on 10 November	Leading

(c) The November 2004 storm (Continued)								
Station	Average value				Peak value			Remarks
	P^a (mb)	T (°C)	H (%)		P^a (mb)	T (°C)	H (%)	
MCM	981.49	-10.84	52.42		984.60 mb at 15:00 UT on 8 November	-8.40°C at 03:00 UT on 9 November	77.00% at 15:00 UT on 10 November	Lagging
Station difference	0.65	-1.47	7.44		1.23 mb (0h)[b]	-2.62°C (0h)[b]	19.36% (0h)[b]	SBA = MCM
(d) The May 2005 storm								
SBA	997.05	-23.77	67.48		1006.4 mb at 15:00 UT on 15 May	-16.46°C at 12:00 UT on 16 May	83.70% at 03:00 UT on 17 May	Lagging
MCM	995.11	-22.05	72.86		1004.4 mb at 15:00 UT on 15 May	-14.80°C at 09:00 UT on 17 May	84.00% at 03:00 UT on 17 May	Leading
Station difference	1.94	-1.72	-5.42		2.00 mb (0h)	-1.66°C (-3h)[b]	-0.30% (0h)[b]	SBA lagging MCM

[a] measured value at sea level

[b] intensity and time difference between SBA and MCM stations.

during this time low wind was recorded, 3.75 m/s for SBA and 13.0 m/s for MCM respectively. For the storm period, the wind profiles at both stations showed moderate activities compared with times before and after the occurrence of storm, even though winds at UT mid-night on 16 May suddenly changed from North-Northeast to North-Northwest directions. After the storms, the wind speed began to increase and reached a maximum value on 18 May, while surface pressure, temperature and humidity showed a similar character to before the storms.

The other surface meteorological conditions (*P*, *T* and *H*) during the period of four major storms of 2003-2005 are summarized in Table 5. In addition, the average local wind speed during the October 2003 storm period at SBA and MCM was about 5.09 m/s and 2.80 m/s, respectively. The local wind speed and wind direction for the November 2003 storm show a similar pattern as was observed in the October 2003 storm. However the event on SBA led MCM by about 1 hour. In the November 2004 storm, the surface pressure and temperature at SBA and MCM show oscillatory changes, humidity decreases to 41% and 29% and increases again before the storm finishes, respectively. The surface meteorological conditions for the May 2005 storm are similar to the November 2004 storm, except the wind was from the Northeast and the East with little variation in speed. During this period, the average surface pressures reached a maximum value of 1005.4 mbar, coinciding with a minimum value of *Dst* (PD). Overall, from day-to-day variation of the local surface meteorological conditions at both MCM and SBA stations, it can be observed that their patterns are almost similar, since these two stations are close to each other, although their topography is quite different.

6.2. The Relation between Solar Activity and PWV During Geomagnetic Storms

The relationship between %ΔTEC and PWV, the global estimated geomagnetic indices (*Kp*) and the local surface meteorological measurements (wind speed and relative humidity) at SBA and MCM are investigated to determine their influence on PWV response. The analyses are presented as follows:

6.2.1. The October 2003 Storm

Figure 17 shows the time series of %ΔTEC and PWV determined at SBA and MCM between 27 October and 2 November 2003 together with the mean %ΔTEC and PWV for the two stations for the same time period. The hourly temporal behavior of %ΔTEC and 3 hours average of PWV profiles for both stations during the observation period are almost similar the maximum peaks of %ΔTEC and PWV are about 362% and 7.43 mm respectively and the standard deviations are 89% and 0.96 mm respectively. The %ΔTEC shows with alternate patterns of positive values (positive phases of the storm) and negative values (negative phases of the storm) with the minimum and maximum %ΔTEC during observation period are -58% and 362% respectively. The PWV profile at MCM led SBA by about 2 hours; however they exhibit a similar trend. The average PWV values at SBA and MCM were 5.24 mm 5.33 mm respectively. Prior to the storm SSC_1, during UT afternoon of 28 October, %ΔTEC was observed to vary between 50% and 100%. A few hours before the onset of the storm SSC_1 until about a few hours after SSC_2, %ΔTEC fluctuated by about ± 30%. At around 21:00 UT on 29 October until 03:00 UT on 30 October a positive phase storm (TEC enhancement) with maximum %ΔTEC of 180% (at 23:50 UT, a few minutes earlier than the PD_1) was observed, this was followed by a negative phase (TEC depletion) of 50% for about 9 hours before the %ΔTEC begin to increase again around 14:00 UT. At the onset of the SSC_2, %ΔTEC increased abruptly and reached a maximum value of 362% at 22:00 UT. On the beginning of UT day 31 October, %ΔTEC began to deplete by about 51% for about 10 hours, during this time GPS data at MCM was not available. On 1 November, around UT noon, %ΔTEC at SBA and MCM show opposite responses (MCM leading SBA) but with nearly similar PWV responses. This could possibly be due to topographic differences between the two stations. Based on the proposed method introduced in Section 5, six distinct episodes of increasing and decreasing PWV, which correlate well with the positive and negative phases of %ΔTEC, except when strong wind occurs at ground level, influencing the distribution of water vapor (increasing or decreasing PWV on a local scale) were observed. These episodes are labeled in Fig. 17 together with solar activity occurrences.

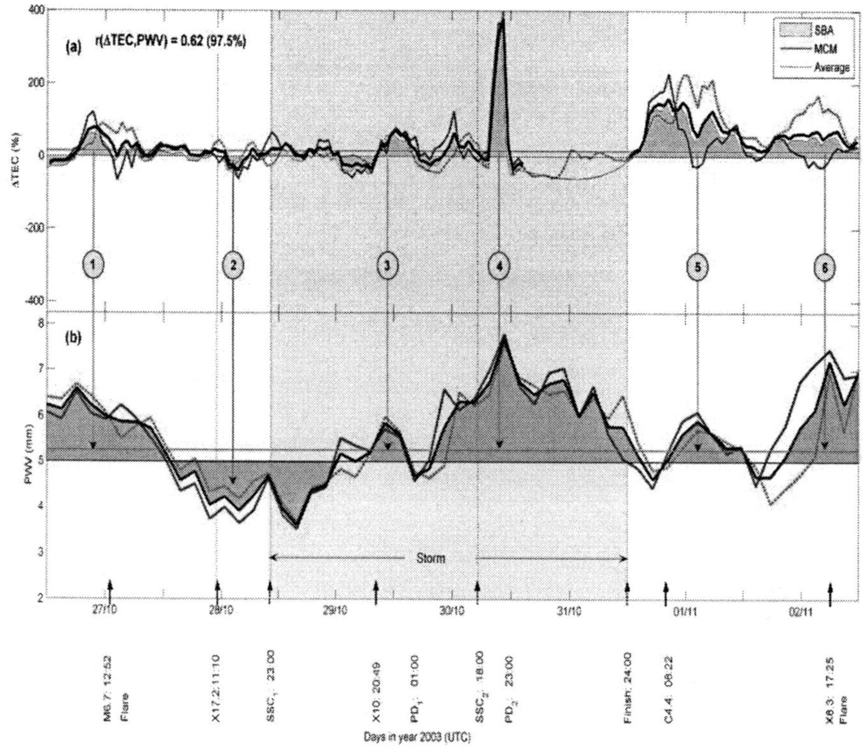

Figure 17. The time series of %ΔTEC and PWV between 27 October and 2 November 2003. The dashed horizontal line is the error margins of %ΔTEC and PWV, respectively. The labels 1 to 6 show the episode of PWV as indicator influence of solar activity through %ΔTEC and the arrow up marks at the bottom of the figure are important time events for storm and flare activity [30].

Referring to Fig. 17 during the first episode of PWV, the geomagnetic storm condition changed from the quiet state to a minor storm state with a combination of C-class and M-class flares, %ΔTEC in positive phase. On the ground relative humidity was high and PWV was observed to increase by about 1.65 mm with respect to PWV_{th}. During the second episode, PWV dropped by about 1.64 mm, and during this period, the storm activity was recorded as active to minor storm, %ΔTEC in negative phase, and the solar flare was high with class X17.2. On the ground, there was a strong local wind with maximum speed of 9.30 m/s from the Northeast. The drop of PWV by about 1.64 mm observed during this period seems to be dominated by a strong and dry local wind even though there was a strong solar flare. During the third episode, the activity of the storm was quite severe with maximum *Kp* equal to 9 and *Dst* minima drop to -350 nT. During this period, an X10 class flare was recorded, %ΔTEC was in low positive phase and PWV increased by about 1.59 mm, lower than that of the quiet to minor storm event of the first episode. During the fourth episode, the activity of the storm was still severe with *Dst* minima equal to -383 nT, and the flare activity was low with C-class flares. During this period, PWV reach a maximum average value of 7.69 mm, an increase of about 2.68 mm and is possibly accompanied by increasing relative humidity. At the fifth episode, the geomagnetic and flare activity are low; the storm began to subside during this time. For this episode, PWV increased by about 1.64 mm and %ΔTEC was in high positive phase with an average value of about 155%. For the last episode, PWV increased by about 1.95 mm. During this period, the storm activity was recorded as quiet, the solar flare with X8.3 class was observed and the surface wind was calm. In this episode, the coupling of the solar flare with the atmosphere is believed to be the major contributor to the increase in PWV. Table 6 gives the summary of these episodes together with the introduction of a term $\pm\delta_{PWV}$ to indicate the increase or decrease in PWV during each episode.

Figure 18 shows the scatterplot and linear regression of PWV versus *Kp*, %ΔTEC, local wind speed and relative humidity for SBA and MCM during the observation period. In the figures, the relation between PWV and *Kp*, PWV and %ΔTEC, PWV and local wind speed (*W*), and PWV and relative humidity (*H*) are obtained based on $Kp \geq 3$, absolute TEC values which exclude the anomaly peaks, a wind threshold (W_{th}) $\geq$ 5.0 m/s and $H \geq 50\%$ which was chosen as it represents the drier conditions that are typically experienced in Antarctic regions, respectively. The PWV increased linearly with respect to *Kp* and %ΔTEC, while PWV decreased linearly with an increase in local wind speed and relative humidity.

Table 6. Summary of PWV episodes during the solar activity from 27 October to 2 November 2003 for PWV event > PWV_{th} [30]

Episode	Time	Description	δ_{PWV} (mm)[a]
1	Begin from 00:00 UT to 23:00 UT on 27 October	**Positive phase of %ΔTEC** Quiet geomagnetic storm C and M class flares Low wind speed Increase humidity (average: 65.3%)[b]	1.65 ± 0.38 (27%)
2	Between 23:00 UT on 27 October and 14:00 UT on 29 October	**Negative phase of %ΔTEC** Active to minor geomagnetic storm X17.2 class flares **Strong wind condition** (max speed: 9.3 m/s)	-1.64 ± 0.57 (-32.8%)
3	15:00 UT to 24:00 UT on 30 October	**Positive phase of %ΔTEC** **Severe geomagnetic storm** (Kp = 9, Dst < -300 nT) X10 class flares **Strong wind condition** (max speed: 10.6 m/s)	1.59 ± 0.46 (31.8%)
4	00:00 UT on 30 Oct to 22:00 UT on 31 October	**Positive phase of %ΔTEC** **Severe geomagnetic storm** (Kp = 9, Dst = -383 nT) C class flares Moderate wind condition **High humidity** (max average: 77.7%)	2.68 ± 0.83 (53.6%)
5	05:00 UT to 21:00 UT on 1 November	**Positive phase of %ΔTEC** Quiet geomagnetic storm activity Low flares activity	1.64 ± 0.38 (12.9%)
6	Begin from 05:00 UT on 2 November	**Positive phase of %ΔTEC** Quiet storm activity X8.3 class flares Calm wind condition **High humidity** (max average: 72.8%)	1.95 ± 0.77 (39%)
	Average positive δ_{PWV} during the observation period		1.84 ± 0.52 (36.8%)

[a] %δ_{PWV} is calculated with respect to 5 mm threshold value.
[b] averages value for SBA and MCM.

The increases in PWV with respect to Kp and relative humidity are quite small in comparison to the significant decrease in PWV with local wind flow. Referring to the linear relationship in Figs. 18a and b and the episodes of increasing and decreasing PWV as shown in Fig. 17, evidence is seen of solar forcing on PWV. As shown in Fig. 18b the high occurrences of PWV are concentrated within the region ±50% of %ΔTEC and PWV varies about ±2 mm from PWV_{th}. PWV was increased due to changes in pressure and temperature although humidity is low as seen in Fig. 18d, this may be attributed to the driest conditions at both stations.

The strong effect of local wind flow on decreasing the PWV in Fig. 18c may be due to the kinetic energy generated by the wind, which has the upper hand on the distribution of PWV in contrast to the solar activity coupling which depends on the intensity of the effect. However as shown in Fig. 19, a strong wind at ground level for SBA and MCM for the period from 22 to 28 May 2003 and for the period from 19 to 25 November 2003 during a quiet solar condition can increase and decrease the PWV, respectively.

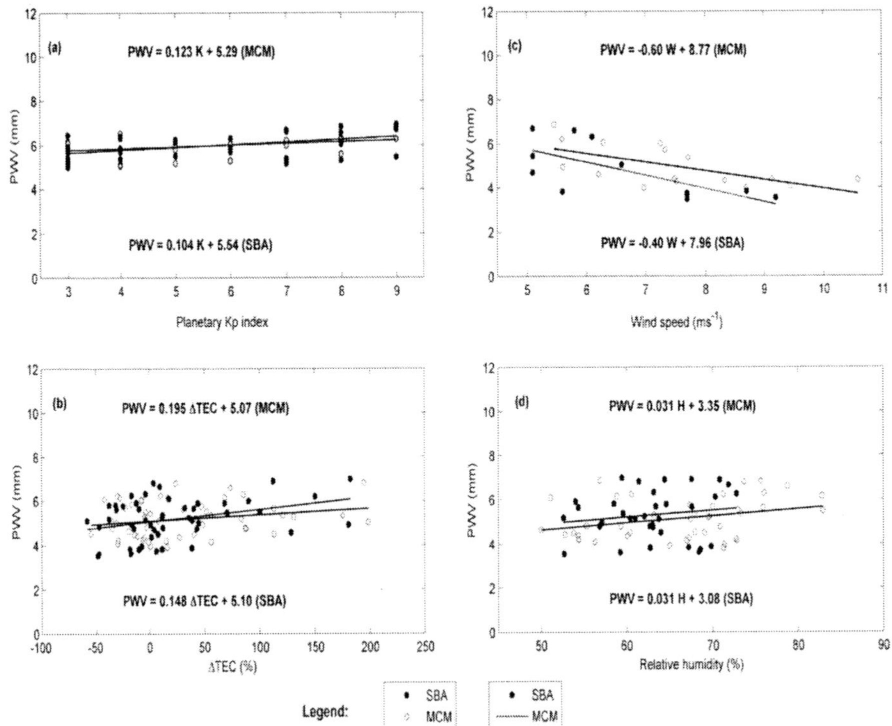

Figure 18. Relationship between planetary (Kp) index, %ΔTEC, local wind speed (W) and relative humidity with respect to PWV from 27 October to 2 November 2003, based on (a) $Kp \geq 3$, (b) absolute TEC values excluding the extreme peaks, (c) the local wind threshold (W_{th}) ≥ 5 m/s and (d) the relative humidity (H) $\geq 50\%$. The equations at the top and bottom in each graph show the linear relation for the observed PWV at SBA and MCM respectively [30].

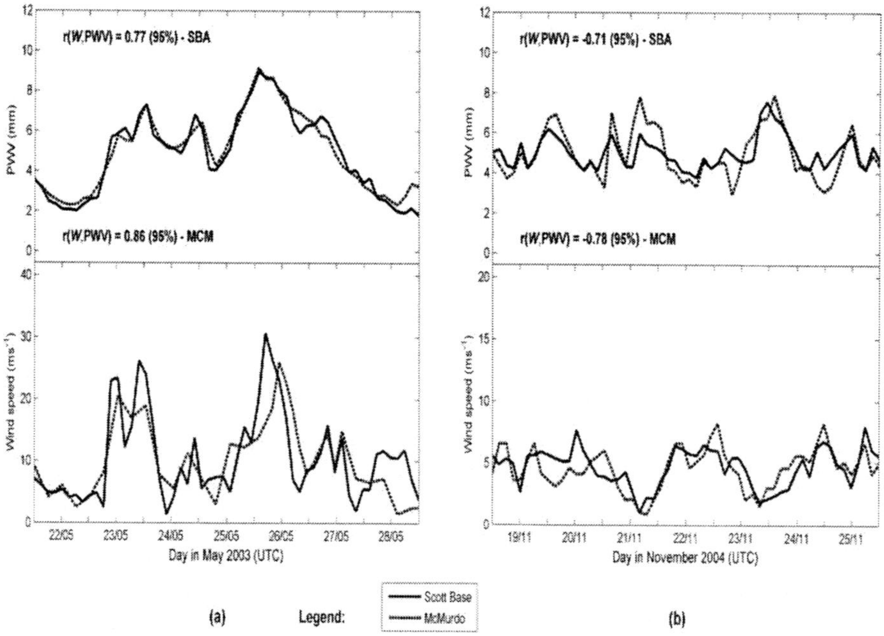

Figure 19. Time series of PWV and local wind speed (a) from 22 to 28 May 2003 and (b) from 19 to 25 November 2004 in 3-hour averages to show the increasing and decreasing of PWV by coupling the local wind speed activity at SBA and MCM respectively. The labels on the graph show the correlation coefficients for the relation between PWV and the local wind speed. Note that the wind flow condition for the event in (b) is smaller by about a factor of 3 than the event in (a) [30].

6.2.2. The November 2003 Storm

Figure 20 shows the time series of %ΔTEC and PWV determined at SBA and MCM between 17 and 23 November 2003 together with their mean values. The %ΔTEC and PWV profiles for these stations had maximum values for %ΔTEC equal to 87.4% and 98.5% and for PWV equal to 7.36 mm and 7.05 mm respectively. The standard deviations of %ΔTEC and PWV were about 31% and 0.78 mm respectively. The hourly temporal behavior of %ΔTEC of the November 2003 storm has longer negative phases than the October 2003 storm. These negative phases have been attributed to changes in the atomic to molecular neutral density ratio, decrease in ionization density and are almost always observed at high latitudes in the summer hemisphere, e.g. [81]. This negative phase is better developed and penetrates to considerably lower latitudes than in the winter hemisphere, which caused seasonal differences in the background wind system [80]. The time series of 3-hour averages of PWV profiles show increasing and decreasing PWV with the PWV profiles at MCM leading SBA by about 3 hours. The average PWV observed during the period were 4.82 mm and 4.78 mm for SBA and MCM respectively.

Referring to Fig. 20a, two strong negative %ΔTEC peaks were observed at UT mid-day on 17 and 20 November and five strong maximum %ΔTEC peaks were observed on 18 and 20 to 23 November respectively. Prior to the storm event a moderate flare with M4.2 class was observed during UT afternoon of 17 November and possibly influenced the drop of %ΔTEC to 25%. After the start of the SSC on 20 November, %ΔTEC dipped below 0% to a minimum average value of -47%. At around 17:30 UT to 18:30 UT (~ 1 hour) on the same day, a short positive phase occurred with maximum %ΔTEC of 83%. This is followed by a long negative phase for about 20 hours before the %ΔTEC began to increase again between 16:30 UT to 22:30 UT on 21 November with maximum %ΔTEC of 60%. This second positive peak could be attributed to a shock by an M9.6/2B X-ray flare in AR 0501 as observed by the CELIAS/SEM sensor onboard the SOHO spacecraft. This flare with a full halo CME occurred at 07:47 UT on 20 November; the following shock occurs 50 hours later on 22 November at 09:59 UT [73].

Using a similar argument as in the October 2003 storm, six distinct episodes of increasing and decreasing PWV which correlate well with the increasing and decreasing of %ΔTEC, were observed. Referring to Fig. 20 and with same explanation as the October 2003 storm, the results for the influence of solar related events on PWV during each episode are summarized in Table 7 together with both solar activity and local conditions marked by highlighting parameters.

Figure 21 shows the relationship between PWV and *Kp*, PWV and %ΔTEC, PWV and local wind speed, and PWV and relative humidity for SBA and MCM. By using a similar threshold to the October 2003 storm, it can be observed that PWV at both stations increased linearly with respect to *Kp*, %ΔTEC and humidity, while PWV decreased linearly with an increase in local wind speed. The amount of increasing and decreasing PWV with respect to %ΔTEC is higher by about a factor of 2 (based on average coefficients of linear regression for two stations) than local wind speed. By comparing the linear relationship between PWV and %ΔTEC as shown in Fig. 21b and in Fig. 20, there is a statistically significant correlation between %ΔTEC and PWV at both stations with a correlation coefficient of 0.77 at 95% confidence level. This is higher by about ~5.5% with respect to the October 2003 storm. By comparing the linear relationships in Figs. 21b and 18b, the increasing %ΔTEC relative to PWV for the November 2003 storm was higher by a factor of 2 than the October 2003 storm. As shown in Fig. 21b the high occurrences of PWV were concentrated within the region ±25% of %ΔTEC.

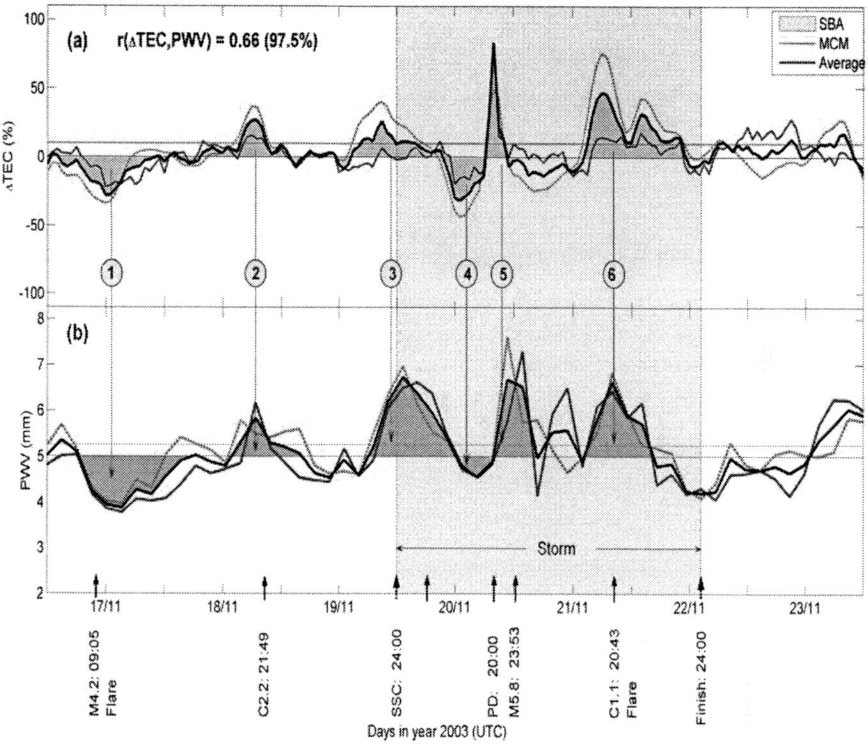

Figure 20. Time series of (a) %ΔTEC and (b) PWV between 17 and 23 November. For details of the labels, see Fig. 17. Note that %ΔTEC for the November 2003 storm is low by a factor of about 2.5 compared to the October 2003 storm [30].

Table 7. Summary of PWV episodes during the solar activity from 17 to 23 November 2003 [30]

Episode	Time	Description	δ_{PWV} (mm)[a]
1	Begin from 06:00 UT on 17 Nov to 06:00 UT on 18 November	**Negative phase of %ΔTEC** **Minor geomagnetic storm** ($Kp \leq 5$, Dst <-40 nT) C and M-class flares Calm local wind speed **Low relative humidity** (min average: 38%)[b]	-1.12 ± 0.44 (-22%)
2	Between 14:00 UT on 18 Nov and 03:00 UT on 19 November	**Positive phase of %ΔTEC** Quiet geomagnetic storm C2.2 and M1.7 class flares Calm wind speed High relative humidity (max average: 75.9%)[b]	0.84 ± 0.34 (17%)
3	Between 16:00 UT on 19 Nov and 13:00 UT on 20 November	**Positive phase of %ΔTEC** **Quiet to strong geomagnetic storm** (Kp = 1 to 7) **Low and moderate flare activity** **High relative humidity** (max average: 81.2%)[b]	1.73 ± 0.55 (34.6%)
4	Between 13:00 UT and 19:00 UT on 20 November	**Negative phase of %ΔTEC** **Strong geomagnetic storm** (Kp = 7 to 8) Low flares activity Plasma decrease (transition time to storm increase) Moderate wind condition (average speed: 5.9 m/s)	-0.45 ± 0.16 (-9.0%)
5	From 19:00 UT on 20 Nov to 01:00 UT on 21 November	**Positive phase of %ΔTEC** **Severe geomagnetic storm** (Kp = 9, Dst < -400 nT) M5.8 class flares **High relative humidity** (max average: 85%)[b]	1.69 ± 0.76 (34%)
6	From 13:00 UT on 21 Nov to 02:00 UT on 22 November	**Positive phase of %ΔTEC** **Quiet geomagnetic storm** (Kp = 3, Dst < -75 nT) Low flares activity Very low wind condition	1.63 ± 0.42 (33%)
	Average positive δ_{PWV} during the observation period		1.47 ± 0.42 (29.5%)

[a] $\%\delta_{PWV}$ is calculated with respect to 5 mm threshold value.
[b] averages value for SBA and MCM.

In Fig. 21c, the influence of local wind on PWV has a similar trend as in the October storm. The strong local wind on 22 November during which the solar activity was calm, was observed to decrease the PWV content. From the relation between PWV and humidity in Fig. 21d, it is observed that PWV at both stations was quite small, increasing at SBA and decreasing at MCM for $H \geq 50\%$. On the other hand, it is interesting to note that the contribution of humidity to the PWV content in this event was rather small, since water vapor only makes up by about ±3% of the total PWV indicating the low amount of moisture in the atmosphere.

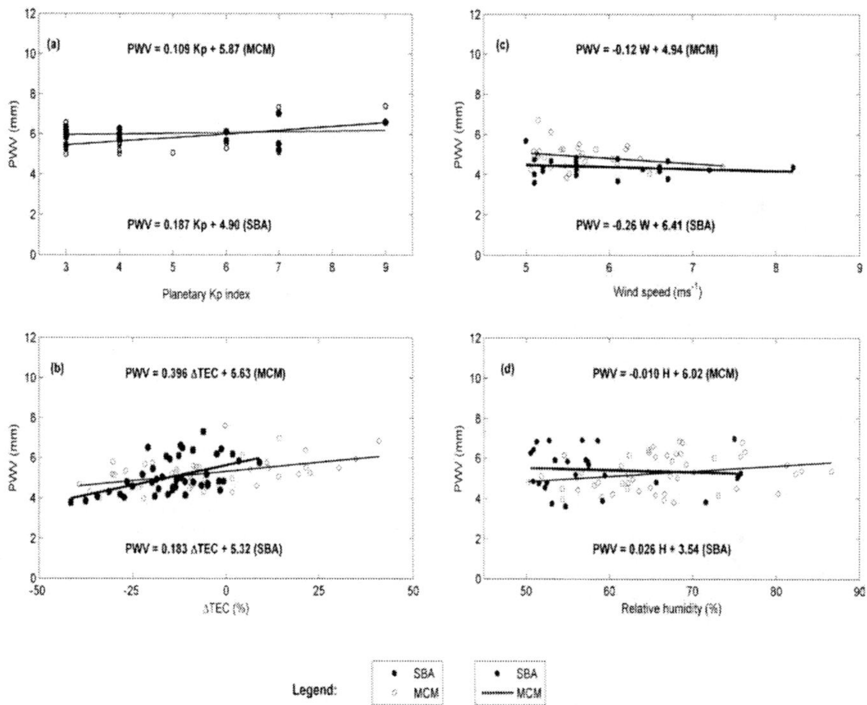

Figure 21. Relationship between (a) estimated planetary (*Kp*) index, (b) %ΔTEC, (c) local wind speed (*W*) and (d) relative humidity with respect to PWV observed from 17 to 23 November 2003 [30].

6.2.3. The November 2004 Storm

Figure 22 shows the time series of %ΔTEC and PWV at SBA and MCM between 5 and 12 November 2004 together with their mean values. The maximum values for %ΔTEC equal to 171.5% and 128.9% and for PWV equal to 8.79 mm and 8.62 mm for SBA and MCM respectively. The standard deviations of %ΔTEC and PWV at both stations were about 43.8% and 1.06 mm respectively. The hourly temporal behavior of %ΔTEC of the November 2004 storm shows longer positive phases than the November 2003 storm. The average PWV observed during the period was 5.67 mm and 5.53 mm for SBA and MCM respectively.

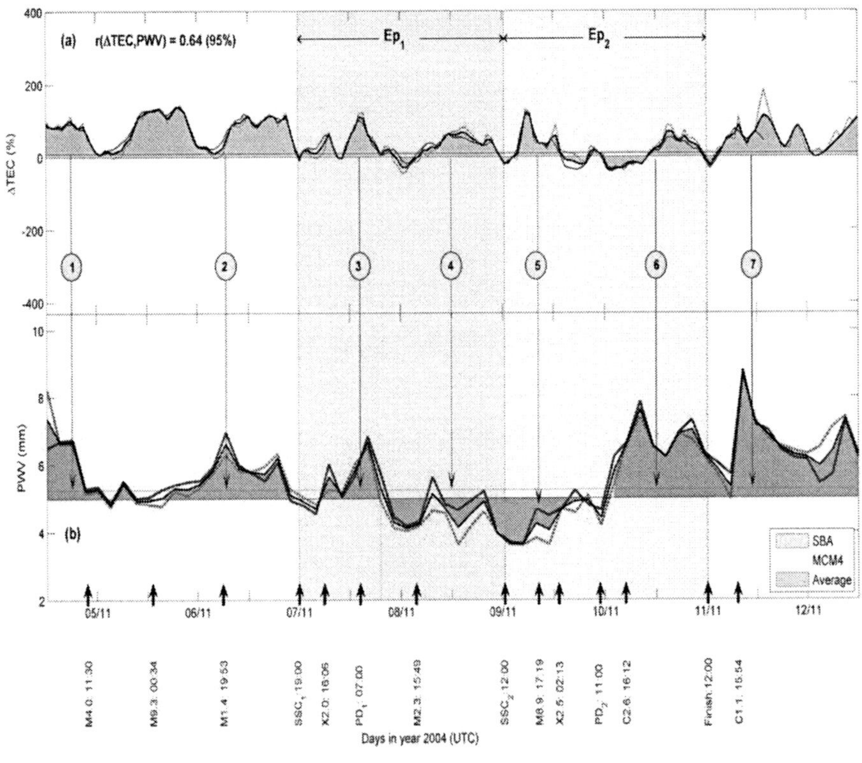

Figure 22. Time series of (a) %ΔTEC and (b) PWV between 5 and 12 November 2004.

Referring to Fig. 22a, eight strong maxima of %ΔTEC and three minima of %ΔTEC were observed during the observation period. Prior to the storm event a moderate flare with M4 class was observed during UT mid-day of 5

November and possibly influenced the increased of +%ΔTEC to 91.4%. Two peaks of +%ΔTEC with maximum average values of 113% and 99% respectively occurred on 6 and 7 November and were possibly influenced by M9.3 class and M1.4 class flares. At SSC_1 on mid-day 7 November, %ΔTEC dipped below 0% to a minimum value of -15%. At around Ep_1, three maximum and one minimum peaks of %ΔTEC were observed around mid-day on 7 and 9 November respectively. The second peak of %ΔTEC occurred with a maximum value of 99.9%, possibly connected with ionospheric variability and storms represented by the first *Dst* peak (PD_1) at 07:00 on 8 November. These events are believed to have originated and been driven by an X2 class flare as well with a full halo CME at the first peak of %ΔTEC (a short positive phase) which occurred with a value of 53.8%, leading to a shock 12 hours later on 9 November [73]. At SSC_2 (i.e. ending of Ep_1), %ΔTEC again dipped below 0%, similar to SSC_1 to a minimum value of -30%. Two strong positive and two negative peaks of %ΔTEC were observed during Ep_2, covering the period from mid-day of 9 to mid-day of 11 November respectively. The first positive peak of %ΔTEC with value of 112.75% could be attributed to an M8.9 class flare. The last positive phase of %ΔTEC in Ep_2 is possibly due to an X2.5 flare and about 8 hours later, %ΔTEC reached a maximum of 58.5%. It is quite obvious that during storm events (Ep_1 and Ep_2), the %ΔTEC values were much lower than before and after the storm, while at the same time the *Dst* index (see Fig. 14) was also the lowest (minimum) in both value and profiles. The anomaly peak during the November 2004 storm was possibly due to the fact that the excursions of IMF *Bz* to North and South were frequent on 7 and 9 November.

To quantify the variation of PWV with solar related events, we observed seven distinct episodes of increasing and decreasing PWV based on the argument introduced in the October 2003 storm. By referring Fig. 22, the summary of PWV response to the solar activities (%ΔTEC, geomagnetic activity and solar flares) and local surface conditions on the ground during of each episode are presented in Table 8.

Table 8. Summary of PWV episodes during the solar activity from 5 to 12 November 2004 [30]

Episode	Time	Description	δ_{PWV} (mm)[a]
1	Begin from 00:00 UT to 12:00 UT on 5 November	**Positive phase of %ΔTEC** Quiet geomagnetic storm **Moderate flare activity** (M4.0 class) Moderate local wind speed	2.36 ± 0.95 (47.2%)
2	Between 22:00 UT on 5 Nov and 12:00 UT on 7 November	**Positive phase of %ΔTEC** Quiet geomagnetic storm **Moderate flare activity** (M1.4 and M9.3 class) Strong wind speed **(average speed: 12.05 m/s)** **Highest humidity** (max: 90.4% (SBA), 84% (MCM))	1.63 ± 0.50 (32.6%)
3	Between 12:00 UT on 7 Nov and 10:00 UT on 8 November	**Positive phase of %ΔTEC** **Unsettled to severe geomagnetic storm** (Kp = 3 to 7) **Strong flare activity** (X2 class flares) Moderate wind speed	1.67 ± 0.67 (33.4%)
4	Between 11:00 UT on 8 Nov and 12:00 UT on 9 November	**Positive phase of %ΔTEC** **Extreme geomagnetic storm** (Kp = 9, Dst = -373 nT) **Moderate flare activity** (M2.3 class) Moderate wind condition (average speed: 8.11 m/s)	-1.04 ± 0.38 (-20.8%)
5	From 12:00 UT on 9 Nov to 09:00 UT on 10 November	**Positive phase of %ΔTEC** **Unsettled to severe** (Kp = 8, Dst < -200 nT) **Strong flare activity** (M8.9 and X2.5 class flares) Moderate wind speed (average speed: 6.32 m/s) **Low humidity** (min average: 35.7%)	-1.37 ± 0.55 (-27.4%)
6	From 14:00 UT on 10 Nov to 13:00 UT on 11 November	**Positive phase of %ΔTEC** **Minor to strong** (Kp = 4 to 7, Dst < -200 nT) Low flare activity **Strong wind condition** (max speed: 35.00 m/s) **High humidity** (max average: 80.5%)	2.63 ± 0.90 (52.6%)

Episode	Time	Description	δPWV (mm)a
7	From 13:00 UT on 11 Nov to 12:00 UT on 12 November	**Positive phase of %ΔTEC** Quiet geomagnetic storm Low flare activity **Strong wind condition** (max speed: 31.00 m/s) **High humidity** (max average: 78.25%)	3.70 ± 0.87 (74%)
	Average positive δ_{PWV} during the observation period		2.40 ± 0.85 (48%)

[a] $\%\delta_{PWV}$ is calculated with respect to 5 mm threshold value.

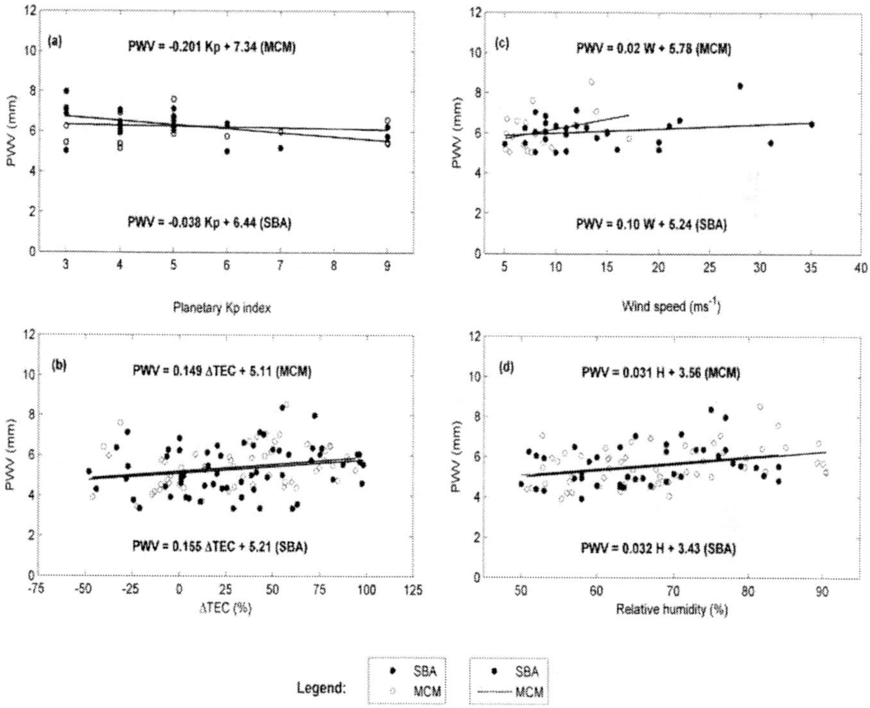

Figure 23. Relationship between (a) estimated planetary (*Kp*) index, (b) %ΔTEC, (c) local wind speed (*W*) and (d) relative humidity with respect to PWV observed from 5 to 12 November 2004 [30].

Referring to Fig. 23, using similar criteria as introduced in the October 2003 storm, it can be seen that PWV decreased linearly with respect to *Kp* at both stations, while PWV increased linearly with an increase in %ΔTEC, local wind speed and relative humidity. PWV decreased with increase *Kp*, contrasting with an increase in %ΔTEC. This event is unique when ionospheric variability represented by TEC alone decreased when *Kp* reached a maximum. PWV consistently increased with increasing %ΔTEC. Their relationships also correlate well between the positive and negative phases of %ΔTEC with increasing and decreasing PWV as shown in Fig. 22, with a correlation coefficient equal to 0.64 at 95% confidence level. It shows a significant increase in comparison with an increase in wind speed and relative humidity. As shown in both Figs. 23c and d, wind and humidity have a strong effect on the PWV. On the other hand, during the observation period the winds and humidity with intense mixing make a significant contribution to increasing PWV, particularly during the second, sixth and seventh episodes. Before and after the storms the wind speed at both stations increased the PWV, opposite to the two previous storms.

6.2.4. The May 2005 Storm

Figure 24 presents the time series of %ΔTEC and PWV determined at SBA and MCM between 13 and 18 may 2005 together with their mean values. The %ΔTEC and PWV profiles for both stations have maximum values for %ΔTEC equal to 172.2% and 209.5% and for PWV equal to 6.22 mm and 5.93 mm respectively. The standard deviations of %ΔTEC and PWV were 47.2% and 0.81 mm respectively. The positive storm phase was dominant during the May 2005 storm, compared with the November 2004 storm, and one strong maximum peak has been observed during the observation period. The time series of 3-hour average PWV profiles show fluctuations of increasing and decreasing PWV and exhibit similar trends. The average PWV observed during the period was 4.43 mm and 4.36 mm for SBA and MCM respectively. Note that before the storm event, PWV increased and decreased; during the storm, PWV gradually increased during the recovery phase and peaked when storm almost subsided; after the storm, PWV begin to decrease.

Referring to Fig. 24a, the mean %ΔTEC fluctuated between –77.55% and 190.86%. Prior to the storm event, two positive phases of %ΔTEC occurred on 13 and 14 May. The two peaks were observed during UT morning on 13 May and UT noon on 14 May with an average maximum value equal to 47.6% and 34.6% respectively. Both increases in %ΔTEC were possibly influenced by M8.0 class and C4.0 class flares, and M8.0 X-ray flare in AR759 with full halo

CME, leading to a shock about 65 hours later on 15 May at 02:10 UT. At around the SSC on 15 May, %ΔTEC started to increase, from 24.2% to 59%. The maximum of %ΔTEC was observed just before the geomagnetic storm with *Dst* peak (PD) equal to –263 nT at 09:00 on 15 May. These storm events were accompanied by a combination of C and M class flares. After PD, %ΔTEC began to decrease and fluctuated between –49.87% (depletion) and 22.2% (enhancement) till the storm subsided. By comparing the *Dst* index of Fig. 15a with the %ΔTEC of Fig. 24a, it can be seen that coupling between TEC and *Dst* is a clear response of ionospheric variability to the geomagnetic storm. Both TEC and *Dst* also correspond well with respect to IMF *Bz* (see Fig. 15), even though the southward turning was lagged by about 3 hours.

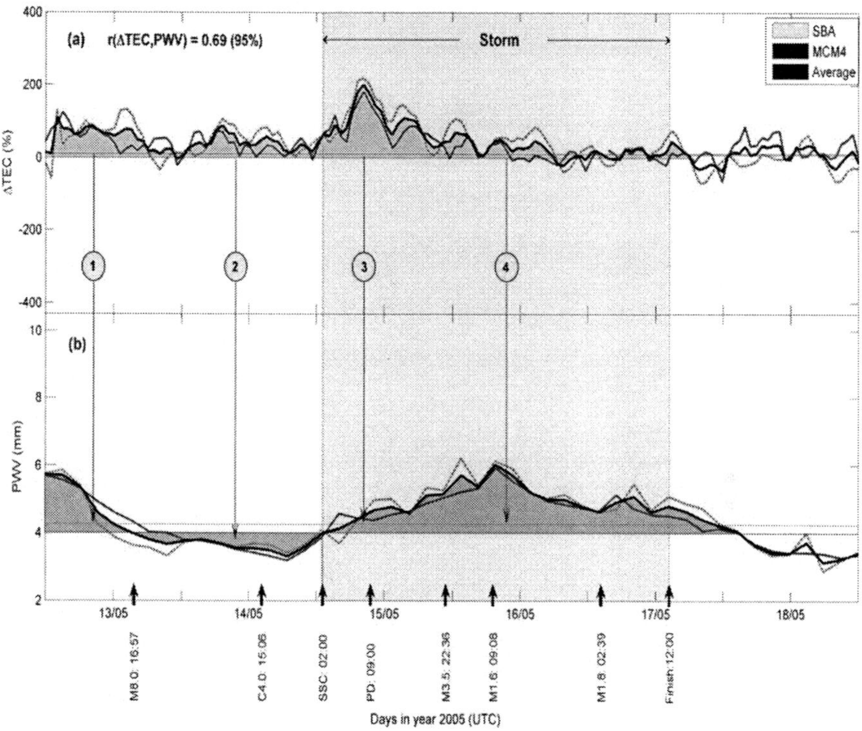

Figure 24. Time series of (a) %ΔTEC and (b) PWV between 13 and 18 May 2005 [30].

Referring to Fig. 24b, the May 2005 storm occurred in late autumn and during this month, the average PWV was 4.05 mm. From PWV data for the year 2005, the average PWV at both SBA and MCM was about 4.46 mm.

Based on this, to show the response of PWV to the TEC, the lower threshold value of PWV (PWV_{th}) was set to 4 mm. In this work, the same error margin as introduced in October 2003 storm is employed. During the observation period, we observed four distinct episodes of increasing and decreasing PWV respect to PWV_{th}. Table 9 gives the summary of PWV behavior during these episodes.

Table 9. Summary of PWV episodes during the solar activity from 13 to 18 May 2005 [30]

Episode	Time	Description	δ_{PWV} (mm)[a]
1	Begin from 00:00 UT to 17:00 UT on 13 May	**Positive phase of %ΔTEC** **Moderate geomagnetic storm** ($Kp > 3$, Dst <-40 nT) **Moderate flare activity** (M8.0 class) **Moderate wind speed at SBA** (average: 6.37 m/s) and **at MCM strong wind speed** (average: 16.6 m/s)	1.85 ± 0.68 (46.25%)
2	Between 17:00 UT on 13 Nov and 01:00 UT on 15 May	**Positive phase of %ΔTEC** Quiet geomagnetic storm Low flare activity (C4.0 class) Low and dry local wind speed **Low humidity and temperature (dry)**	-0.75 ± 0.24 (-18.75%)
3 and 4	Between 02:00 UT on 15 Nov and 02:00 UT on 18 May	**Positive phase of %ΔTEC** **Quiet to severe geomagnetic storm** (Kp = 2 to 9 and Dst = -263 nT) **Moderate flare activity** (M3.5, M1.6 and M1.8 class)	2.08 ± 0.54 (52%)
	Average positive δ_{PWV} during the observation period		1.96 ± 0.17 (49%)

[a] %δ_{PWV} is calculated with respect to 4 mm threshold value.

Figure 25 shows the relationship between PWV and *Kp*, PWV and %ΔTEC, PWV and local wind speed, and PWV and relative humidity. It can be observed that PWV increased linearly with respect to *Kp* and %ΔTEC at both stations, PWV decreased linearly with an increase in local wind speed, while relative humidity showed a small effect, decreasing and increasing the PWV. The increase in PWV with respect to *Kp* is by a factor of 2.95 (on average) and is higher by 4 times in comparison with the three previous geomagnetic storm events. The increase in PWV by %ΔTEC is smaller by about 4 times with respect to *Kp* and higher than relative humidity (on average) and is quite small in comparison with decreasing local wind speed. The strong effect of local wind flow ($W > 5$ m/s) on PWV has a similar trend as in the October 2003 and November 2003 storms. In this event, the strong local wind speed at both stations during the first episode, in which active solar activity ($SSN > 100$ and $F_{10.7} > 100$), was observed to increase the PWV content. Furthermore by comparing the peak of %ΔTEC in Fig. 24a and the PWV peak in Fig. 24b, is seen the relation between the positive and negative phases of %ΔTEC and the amount of increasing and decreasing of PWV is well correlated with a correlation coefficient of 0.69 at 95% confidence level, although there appears to be a time lag for PWV by about 24 hours. The increase in PWV with respect to the TEC measurements was observed consistently with increasing %ΔTEC for all storms in this analysis. This implies statistically significant solar forcing on PWV.

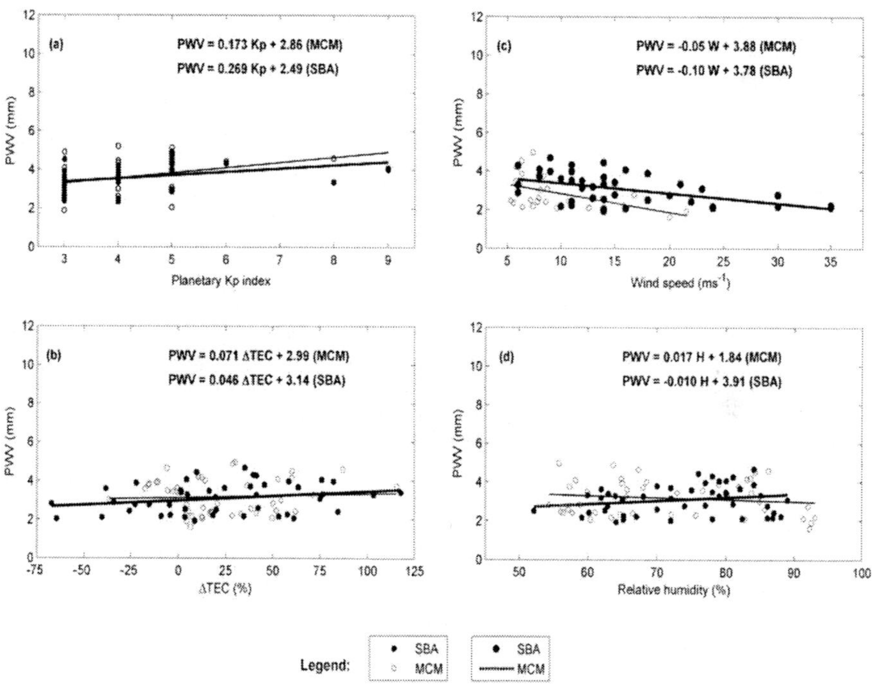

Figure 25. Relationship between (a) estimated planetary (Kp) index, (b) %ΔTEC, (c) local wind speed (W) and (d) relative humidity with respect to PWV observed from 13 to 18 May 2005 [30].

6.3. Solar Activity and PWV During Quiet Days

Analysis of solar activity and PWV as described previously was observed for the storm period including two days before and after the storm event. However, two days before and after the storms period are not really quiet, so many factors can disturb them. The PWV for quiet days need to be compared with PWV variation during storm events. A quiet day is defined as a day when no storm or a very weak storm occurs. In this section, quiet days based on International quiet (Q)-days from GFZ [82] were employed. Further, the first quiet day or the quietest days (Q1) in each month during the storm events was chosen. This quiet day was also used to determine TEC_{QD} in Eq. (36).

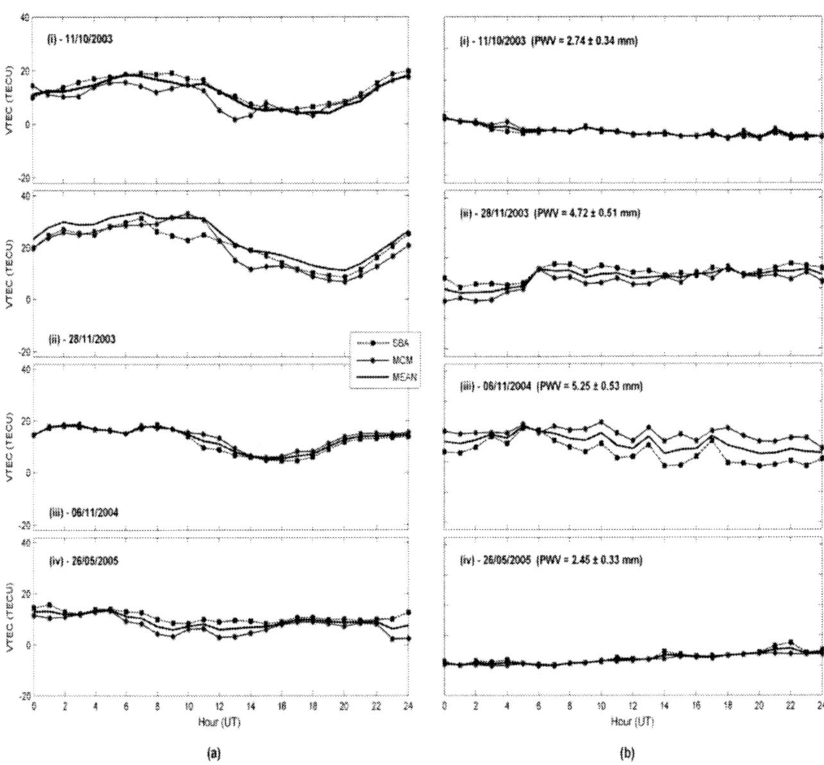

Figure 26. (a) Time plot of TEC and (b) PWV on hourly mean basis for SBA and MCM respectively. From top to bottom shows both TEC and PWV variations during the quiet day of October 2003 storm, November 2003 storm, November 2004 storm and May 2005 storm. At the top each figure shows the quiet day and PWV variation during the day [30].

Figure 26 shows the ionospheric variations of TEC and PWV at SBA and MCM together with their mean values during quiet days. In the top or bottom of each figure are shown the quiet day (Q1) events selected for each geomagnetic storm. In Fig. 26a, TEC as a solar activity parameter exhibits a maximum (TEC enhancement) before noon at around 07:00 - 10:00 UT, decreasing during mid-day UT and reaching a minimum (TEC depletion) at around 15:00 - 22:00 UT. From this figure, it is clear that the intensity of TEC affecting the Earth's atmosphere varies with the diurnal rotation of the earth, with a TEC maximum around 07:00 - 10:00 UT (19:00 - 22:00 LT) similar to the 'Fountain effect" of the equatorial anomaly [71]. In Fig. 26b, the hour-to-hour variation in PWV is quite small and shows an almost similar pattern. By comparing Figs. 26a and b, it can be seen that PWV does not follow the TEC pattern. This indicates that during quiet days the influence of solar activity on PWV is probably by other factor such as advection transport due to local surface conditions. In other words, heat energy from the Sun is not enough to evaporate water (low concentration of water vapor) from the Earth's surface to make evaporation-condensation cycle through the atmosphere [83] and therefore vertical distribution of PWV content is rather difficult detected by GPS measurements.

6.4. The Relationship between Solar Activity and PWV on Monthly Basis

To show significant response of water vapor to solar activity, raw data (all days) of PWV was separated into quiet days and storm-affected days. However, the classification of quiet days from GFZ as introduced in Section 6.3 sometimes was not consistent with terrestrial conditions. To overcome this problem, the assumption of quiet days was redefined. The classification of quiet days from Gonzalez et al. [84] and World Data Center (WDC) were considered. Furthermore, in the solar system, solar flares are also a major cause of geomagnetic disturbances and have a direct effect on the Earth's atmosphere. On the ground, strong local wind speed, particularly Katabatic wind in Antarctica, also plays an important role in the global climate system, e.g. [85, 86] and the winds show an effect on PWV [11, 30]. Based on this assumption, solar flare activity (X-, M- and C-class) and local wind speed (Ws > 5 m/s) are included in determining a storm day together with the storm classification from Gonzalez et al. and WDC.

Figure 27 shows the time plot of solar activity (TEC) and PWV based on monthly averages during 3-year period observation. The calculated PWV at SBA for the period of 2003 - 2005 was less than 10 mm (monthly average). The mean and standard deviation values based on daily averages ranged between 4.35–5.16 mm and 0.12-1.43 mm respectively. The PWV at SBA was lower than at MCM, with PWV mean difference of 0.20 mm. As shown in Fig. 27a, the month-to-month variation of TEC at both stations exhibits a similar pattern, with mean value of TEC during the period of 2003-2005 is ranging between –1.79 and 17.07 TECU. During this period, TEC value at SBA is lowest by about 23.2%, while at MCM it is higher by about 30.3% from their respective mean values. At the same time, TEC between SBA and MCM differs by about ±5 TECU. Obviously, the TEC peaks at both stations show clear evidence of seasonal signal, indicating higher relative ionospheric activity during the summer period (December, January and February, DJF) and lower activity for the winter period (June, July and August, JJA). The summer and winter peak averages of TEC are 12.12 and 0.55 TECU respectively.

Figures 27b-d show the month-to-month variations of PWV at both stations together with their mean values for all, quiet and storm days. On a monthly basis, the variation of PWV (Fig. 27b) from January 2003 to December 2005 ranges from 1.34 to 10.02 mm for SBA and from 1.45 to 10.30 mm for MCM. Referring to Figs. 27b and c, amounts of PWV over the year are almost similar to the TEC, with their mean PWV difference between all days and quiet days was 0.025 mm. The small differences in mean value between quiet and all days are possibly caused by conditions at both stations being driest. Notice that for both stations, the mean PWV show large seasonal variations precipitation, largest in summer and lowest in winter, when temperature and humidity are high and low respectively. This observation is also similar to the GPS estimated PWV as reported by Vey et al. [87] measured over a period from 1997 to 2002 at six stations in Antarctica, less than 10 mm, where McMurdo (MCM) is one of the stations in the Vey et al. analysis. For storm- disturbed days in Fig. 27d, PWV exhibits a maximum of 6.55% compared to PWV in quiet days. PWV variations show a clear seasonal signal, following the TEC variation. The trend in PWV during the storm also varies significantly, which is expected from the TEC trend. However, the PWV has a lag time of about one month behind the TEC variation. It is interesting to note that the monthly mean of GPS PWV variation over the years of 2003-2005, particularly during storm events, show a close correspondence to the solar activity as represented by TEC variation and as presented here, are the first such relation for Antarctica. In addition, the trend in both TEC and PWV signals follow the trend in the sunspot cycle, i.e. descending in the phase of solar cycle 23. Although GPS data for Antarctica are not complete for an eleven year solar cycle, we believe that the observed data contains a solar cycle pattern, as lower TEC values are a real effect during the minimum sunspot cycles.

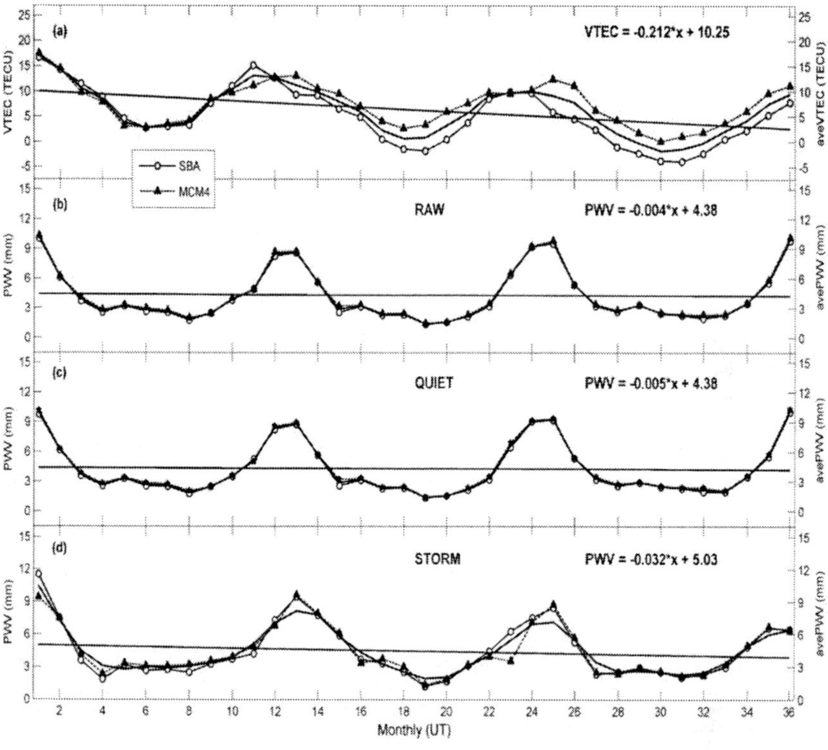

Figure 27. Solar forcing on PWV from January 2003 to December 2005: (a) Time plot of absolute TEC, (b) PWV from all days (raw data), (c) PWV during quiet days and (d) PWV for storm days. The solid line shows the monthly averages from two stations and the straight line is a linear regression to show the trend of solar activities and PWV. Note that TEC and PWV are processed from monthly average. A clear seasonal signal is present for both TEC and PWV [30].

Besides investigation relationship between solar activity and climate at SBA and MCM, three additional stations at Antarctica was also compared. Figure 28 shows the time plot of solar activity (sunspot number, solar flare index (SFI) and TEC) and PWV (for all, quiet and storm days) for the period of 2003-2005. Both SSN and solar flare index (SFI) are plotted and categorized as a direct influence of solar activity on the terrestrial atmosphere. The SFI in Fig. 28a is plotted only for the total solar disk, because SFI variations for northern and southern hemispheres are similar. In this work SFI data for the period from January 2003 to December 2005 was supported by Ataç [88] from Kandilli Observatory, Turkey. It was found that the yearly mean values of SFI in the northern hemisphere, southern hemisphere and total disk surface of the Sun are very closely related to the SSN [89]. Figures 28b-d shows the monthly variability of TEC and PWV for five stations at Antarctica. The monthly mean of TEC variation in each station exhibits a similar pattern and is also corresponding well with its geographical/geomagnetic locations. At SBA, the TEC for the period of 2004-2005 is lower by about 4% from their mean value, while at MCM4, CAS1 and SYOG having intermediate values for all periods. The high value of TEC for the period of 2003 was occurred at MAW1. On the whole, the TEC mean during the period of 2003-2005 is vary between –0.71 and 18.76 TECU (monthly average). The TEC value in Fig. 28b shows clear seasonal signal evidence and each station during summer is higher than those of winter. The summer and winter TEC peaks average for all stations are 13.15 and 2.30 TECU respectively. The TEC also varies annually, with a maximum at the equinoxes. In the next panel of Figs. 29c-e shows the plot of monthly mean of PWV. Comparing Fig. 29c (all days) and Fig. 29d (quiet days), the PWV variation was almost a similar character in signal and trend. By separating the storm-affected days from raw data, it can be shown that when storm occurred both in space and on the ground (see Fig. 29e), PWV variation was highest of about 19.2% compared to the PWV in quiet days. Additionally, in the first summer of 2003, the PWV was reached more than 10 mm.

Relationship between PWV and TEC for raw data (all days), absence of storm (quiet days) and storm affected-days, and for two and five sites respectively is plotted in Figure 29. As shown in the figure, their relationships on monthly averages based show a strong correlation. Referring to Figs. 29a (i and ii), the correlation between monthly mean values of TEC and PWV at both stations for all days is higher only by about 1.5% in comparison with quiet days. However, the correlation coefficient during storm events (Fig. 29a-iii) is largest by about 9.2% (SBA) and 12.7% (MCM) and 10.7% (SBA) and 14.1% (MCM) in comparison with all days and quiet days, respectively. Similar to Figs. 29b (i and ii), the correlation coefficient during storm events ($r = 0.84$) is also largest by about 15% and 9% in comparison with quiet ($r = 0.70$) and all days ($r = 0.73$), respectively.

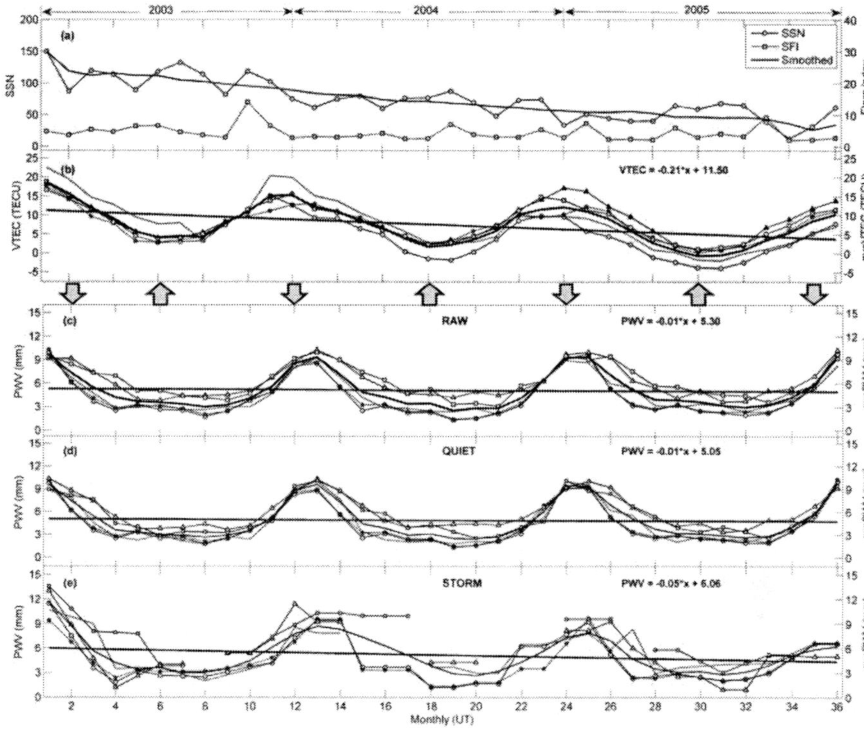

Figure 29. Solar forcing on PWV from January 2003 to December 2005 for five stations in the Antarctic region [48].

The higher correlation during storm days of more than 10% with respect to quiet days and all days suggests water vapor plays a significant role in the high-energy thermodynamics of the atmosphere during the occurrence of storm events. This showing solar activity accompanied by local perturbations of the ambient medium force a water vapor response. The results show a good agreement in relation between PWV and TEC particularly during storm events, and solar activity influencing PWV can be inferred through their relationships. On the other hand, small differences in both conditions indicate that both stations are in dry conditions. SBA and MCM are located on the Antarctic Ross Ice self continent and is very low water vapor content due to recurrent blowing snow compared to other stations. The small difference in correlation between storm and quiet days was occurred at CAS1 due to high humidity and warms the surface temperature at this station. Besides the conditions being driest and huge impacts of surface slope or topographical/geographical conditions, the lowest precipitation amounts spatially at Antarctica are possibly due to lack of synoptic forcing [90]. The elevated terrain acts to block the penetration of synoptic disturbances onto the plateau, when the slope is steep and the large-scale wind is directed upslope, the rising air results in enhanced precipitation. Precipitation increases with slope and latitude and decreases with elevation and continentally. Additionally, blowing snow and strong winds can advect out the precipitating snow from the regions; the passage deep of low-pressure depressions is one of major contributor. King and Turner [91] was also found that the amount of blowing snow across the Antarctic area is dependent on both temperature and wind speed. At higher wind speed and lower temperatures, the snow will be more likely to be advected into a region, while higher temperature and low wind speeds will tend to deter blowing snow. These disturbances can affect the GPS signals. One of these effects is often referred to as the "tropospheric delay", which has been explained in Section 2, since the largest part of the effect occurs in the troposphere caused by water vapor, rain, clouds, snow and other particulates.

From Figs. 27-29, it was observed that solar activity represented indirectly by TEC measurement, on a monthly analysis shows forcing the PWV during storm events. Based on this connection, firstly we note that one of the most prominent peaks in their profiles exhibits a similar pattern of high and low fluctuations at all stations, with a small TEC differences in the intensity and phases, suggesting an annual periodicity was present in the Antarctic region. Secondly, a strong relationship existed between PWV and TEC during storm events with correlation coefficients vary from 0.70 to 0.84 at the 99% confidence levels. These seemed to correlate with a significant seasonal signal in the Antarctic region corresponding to the ionospheric TEC, with highest peak (very active) in summer polar days and lowest peak (inactive) for the winter polar night periods. Finally, during the observation period, noticed that monthly PWV mean for the storm events was shown downward trend, with a slope of -0.05 ± 0.15 mm/day/month, indicates a significant response of the PWV variability to the solar activity. These suggest the temporal distribution and strength of solar activity as indicated by geomagnetic disturbances and ionospheric activity are important inputs for the seasonal cycle of water vapor. Although at this time the GPS data is not complete for one solar cycle on the globe, it is anticipated that the observed data exhibits a similar trends in the variation between TEC and PWV with respect to solar cycle pattern.

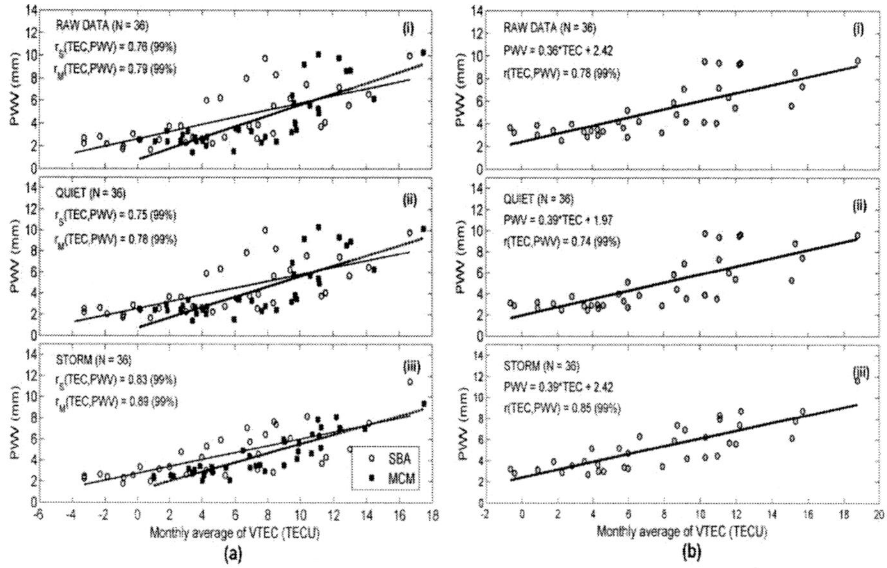

Figure 29. The relationship between monthly mean of TEC and PWV during the period of 2003-2005 for (i) all days, (ii) quiet days and (iii) storm days at (a) SBA and MCM, (b) at five stations. The coefficient correlation (r) between TEC and PWV is calculated at the 99% confidence level and subscripts S and M on the left-side of graph stands for SBA and MCM [30, 48]. Note that the correlation between TEC and PWV at MCM is consistently stronger than that at SBA.

Chapter 7

PROPOSED PHYSICAL MECHANISMS

Changes in the Sun's activity have a profound impact on the Earth's atmosphere and their impact clearly drives the Earth's climate and controls the weather. There is a long history of solar-climate studies [92]. In this study, it was observed that the direct influences of number of sunspot and total solar irradiation variables on atmospheric water vapor for short-term analysis are ambiguous without a clear correlation, even observed through the poles. Thus, the influences of solar activity on terrestrial climate are correlated indirectly using TEC measurements during geomagnetic storms, which were investigated in Sections 4-6. In this part, a physical mechanism between upper and lower levels of the atmosphere is proposed by assuming an intense heliogeophysical event such as a geomagnetic storm has occurred. TEC is used to linkage between the Sun and tropospheric activities. Figure 30 shows the new possible mechanisms of solar influence on the weather and climate through ionospheric-tropospheric coupling.

Referring to Fig. 30, the Sun is the fundamental cause of various disturbances in the near-Earth space environment (geospace). The Sun effects on the Earth's magnetosphere, ionosphere and in the atmospheric dynamics, particularly on the Earth's magnetic field lead to an important event, known as the geomagnetic storm. This tends to occur during time of geophysical disturbances resulting from increase in solar activity (changes in solar irradiance through sunspot and faculae, changes in solar flare frequency and intensity, and changes in the various coronal phenomena) and their energy enter the Earth's magnetosphere communicated via the solar wind [93, 94, 95]. During this event, the Earth's magnetic field acts as wires to relay the energy from solar wind to the ionosphere in the form of heat [96]. The transfer of energy from the Sun to the magnetosphere by the solar wind is controlled by electric fields. The key mechanism in the coupling of the solar wind with the magnetosphere is the reconnection [97]. The coupling sets up a large system of magnetic-aligned currents (FACs). The change of current (energetic particle) during extreme solar activities flowing through the ionosphere can cause disruption to power distribution systems, long-distance telephone networks and corrosion of pipelines on the ground [96].

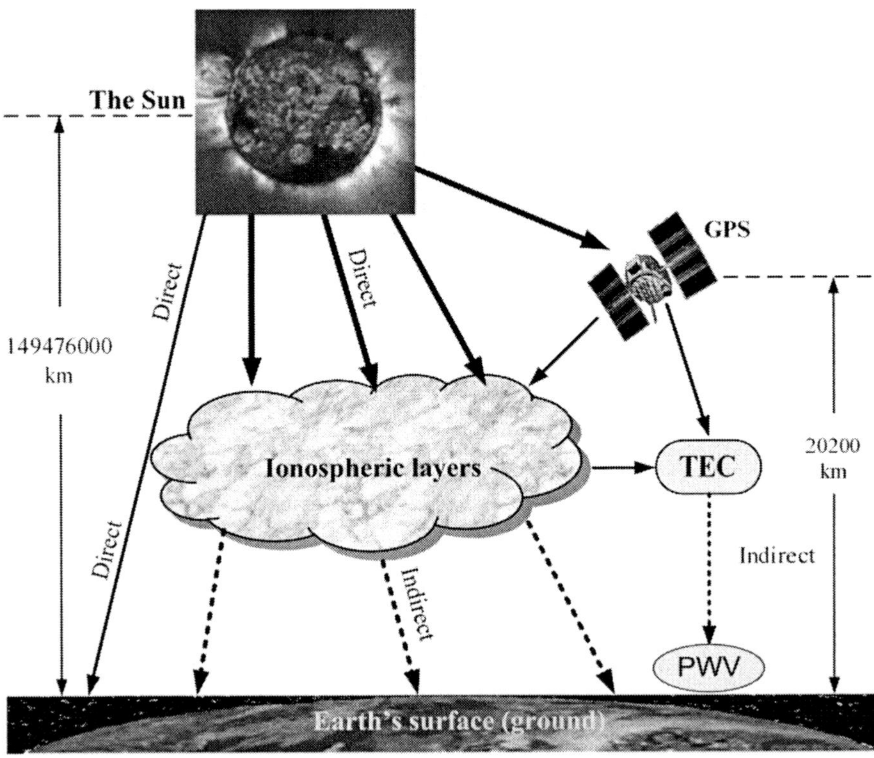

Figure 30. Illustration of the GPS TEC-PWV method [48].

Energetic particle effects become more intense and pose a hazard to astronauts and damage spacecraft electronics; satellite lifetimes are shortened by increased atmospheric drag and the changing plasma environment disrupts communications and navigation [98]. In communications, these disturbances can affect the GPS signals [99]. Thus, one will consider the ionospheric virtual path extension of the signal propagation through the ionosphere measured by the TEC. TEC is the number of electrons in a column a square meter cross-section along the ray path between transmitter and receiver. Solar X-rays and extreme ultraviolet (EUV) radiation in the upper atmosphere are causes the sudden increase of TEC. Their produces free electrons, since the ionization rate is dependent on solar radiation and particles emitted from the Sun [100, 101]. This disturbance is possibly also acting as a mechanism in the ionosphere-troposphere coupling.

Diurnal variation of TEC at any specific region is clear representing the amount solar radiation in that region. The TEC variation along a signal path depends on the ionization and recombination of particles as well as on the transport processes in the ionosphere. The recombination process is dependent on the collision rate between the particles, and is thereby a function of density, pressure and temperature. Both ionization and recombination processes are occurred during dayside (when the Sun activity is high) and night-time (when the Sun activity is low), respectively. The combination of ionization and recombination processes forming a diurnal cycle in the TEC variation. From both transport processes, a diurnal cycle of TEC has a maximum at 14:00 and a minimum at sunrise, local time (LT). On the other hand, the intensity of radiation from the Sun affecting the Earth's atmosphere varies with diurnal rotation of the Earth, so the strongest ionization is occurring on the dayside of the Earth, while the night side is less affected by the radiation. Looking at the specific event, that is during the period of high geomagnetic activities (highest energy), the interaction between the Sun and TEC is well correlated [71]. Their interaction resultant electric fields produce remarkable effects in the ionosphere and the troposphere. The increase in TEC indicates substantial disturbances to the global atmospheric circuit, through an increase in the incident density of energetic particles in the atmosphere, and a major change in the conductivity in and above the ionospheric D-region. The possible ionospheric activity forces the water vapor is therefore believed through electrical discharge from lightning-induced electron precipitation events could propagate downwards from clouds to the troposphere. These effects are expected to be greater at high latitudes due to the enhanced incident particle transport along geomagnetic field lines. Further at high latitudes, altitudinal redistribution of condensation nuclei occurs under increased electric field current of the atmosphere. This is modulated by cosmic rays resulting water vapor condenses in area where the concentration of these nuclei is small and there is sufficient. These processes is accompanied by water vapor phase transition and clouds formation. The cloud leads to change in radiation balance, reduced radiative cooling and altered thermo-baric field in the troposphere. The details of how solar activity influence on PWV or troposphere parameter is illustrated in Figure 31.

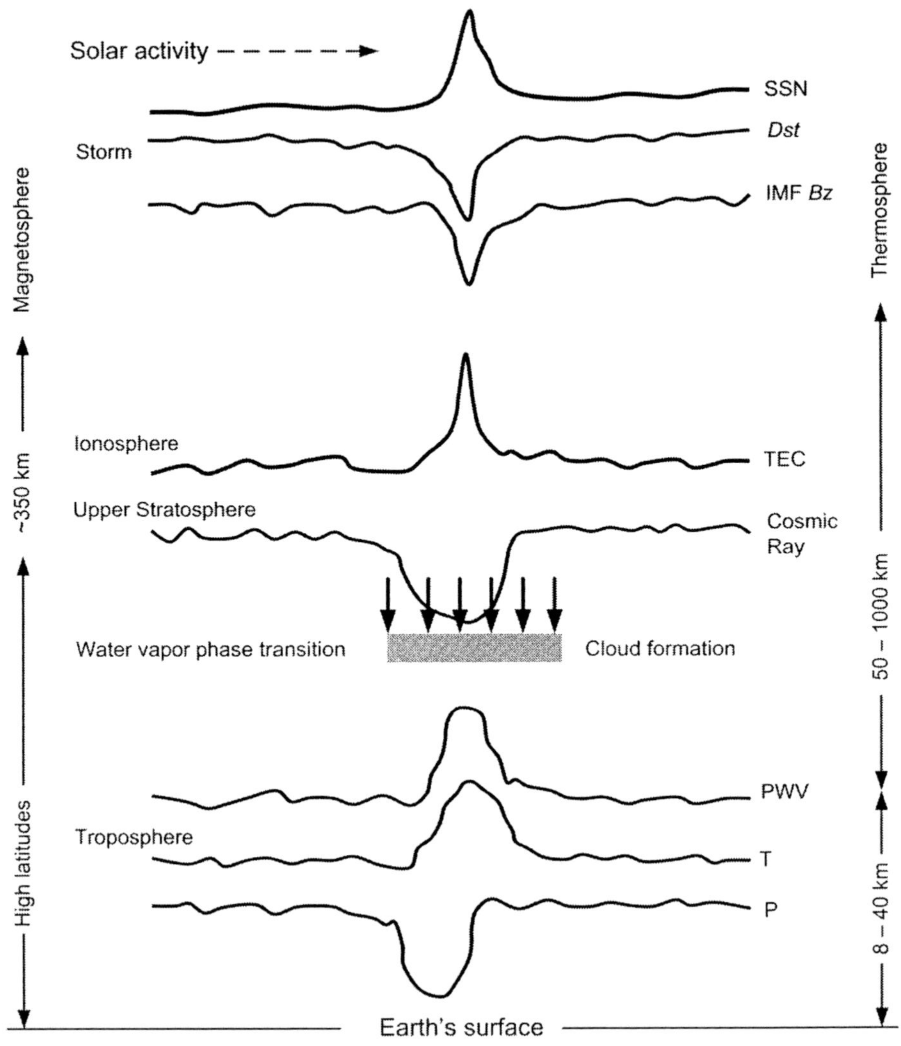

Figure 31. Detail illustration of physical mechanisms of solar influence on PWV during geomagnetic storm (atmosphere layer not to scale) [48].

Manifestation of these effects in the troposphere will depend on the time of day, season and atmospheric condition in a given time and region. For example within this process, it can be seen that the PWV response to the TEC during major geomagnetic storm has a lag of about 4 ~ 6 hours. However, the PWV response on the troposphere through seasonal changes of surface temperature to the TEC was increased during summer and decreased during winter [27, 30, 48]. The surface temperature and pressure were the biggest contributing factors to changing the PWV content in the atmosphere [11, 30]. Both these parameters are susceptible to changes in the global atmospheric electric circuit. Harrison and Ambaum [102] noted a change in cloud base temperature at a middle latitude station during a solar flare. Burns et al. [103] describe statistically significant changes in surface pressure with changes in the vertical electric field at Vostok station in Antarctica. The global atmospheric electric field provides a clear physical link between TEC in the ionosphere and PWV near the surface at high latitudes during the occurrence of intense solar flares. These mechanisms suggest that the temporal distribution and strength of solar activity (shown by TEC) as indicated by geomagnetic disturbances and ionospheric activity are important inputs for seasonal cycle of water vapor. This is one possible relationship between TEC and PWV in the atmosphere and their influences can be inferred from ionospheric disturbances, as indicated by delaying the GPS signal. Notice that the TEC in this approach is not directly controlling the PWV, but TEC is used as an indicator of solar influences on terrestrial climate.

In addition to the TEC-PWV relationship, the TEC varies annually (see Fig. 29), with a maximum at the equinoxes and a minimum during the winter [30, 104] following the solar cycle pattern [27]. Therefore at high latitudes, the winter low occurs because the Sun's radiation no longer create new ions and the recombination process depletes the existing ion supply, hence TEC is depleted. Similar for the summer season, since a higher temperature increases the amount of ionization taking place in the ionosphere, whereby TEC is enhanced. Finally, TEC varies with the 11-year solar cycles, since ionization rate is depending on solar radiation and particles emitted from the Sun. Figure 32 presents the variation of solar activity including flares activity (shown by SFI). Solar cycles are cyclic and fluctuating due to changes in the amount of energy emitted by the Sun. The amount of emission from the Sun (TSI at the surface) is about 1367 Wm^{-2} (on average) and is correlated with the number of sunspots and solar flux. The sunspots and also the solar activity, varies according to a cycle that has an average length of 11 years (*Schwabe cycle*). As shown in the figure, during the periods when the Sun is more active (high), the Sun also emits radiation caused by solar flares. These are sudden eruptions from the Sun, often occurring close to sunspots, causing an intense emission of ultraviolet radiation. Generally, ultraviolet radiation from the Sun, along with radiation from outer cosmic sources, drives ionization process in the ionosphere [99]. This cycle is well known, with four stages: the minimum, rising, maximum and declining phases. In the figure, the period of 2003-2005 was in declining phase of solar cycle 23, when TEC seen decreased. This suggests that the solar activity represented by TEC follows the solar cycles and it can be used as an indicator solar-climate link.

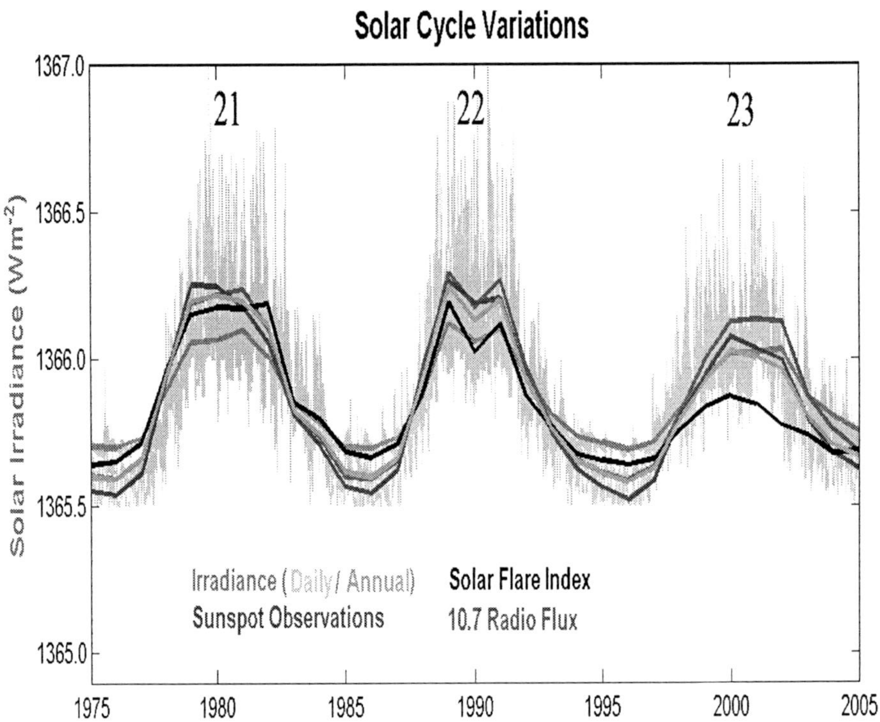

Figure 32. Solar activities (total solar irradiance, sunspot number, solar flare index and solar flux) from 1975 to 2005 [48].

Chapter 8

CONCLUSION

The vertical integrated PWV in the Antarctic region was characterized using the GPS signals and the surface meteorological data. The influence of solar related events (solar flare and geomagnetic storm) on PWV was determined indirectly for the first time based on a novel technique of correlating the GPS PWV and TEC measurements. The advantage of this method, it can be used for climate studies at short-term data with a strong event is required. Based on the analyses of the four intervals of major geomagnetic storms showed that PWV was observed to consistently increase with increasing %ΔTEC for four cases of geomagnetic storms. The %ΔTEC at both SBA and MCM stations exhibit similar patterns, but with small differences in the intensities and phases, which could be attributed to the local differences between the two stations. Their relationship was observed strong with correlation coefficients vary from 0.64 to 0.77 at the 95% confidence levels, and these give statistically significant evidence of solar influence on the PWV. The high increased TEC values in Antarctica were consistent with the days disturbed by storms coming from both space and ground activity respectively. However, the November 2004 storm was unique when TEC was higher before and after the occurrence of storm. Outside the overall storm period, PWV and %ΔTEC still vary and almost follow each other but their correlation is less significant, except during significant solar activity. An examination of the influence of local wind conditions on PWV showed that PWV decreased with an increase in local wind speed. Furthermore, the influence of humidity on PWV was observed to be relatively small. On a monthly analysis, the PWV during storms was observed higher by ~10% in comparison with quiet days, and the correlation between PWV and TEC for a

3-year period of observations was 0.85 (on average) at the 99% confidence levels. On the other hand, solar influence on PWV in the Antarctic region was clear shown a significant seasonal signal, which is very active (higher) during the summer and inactive (lower) for the winter periods.

Understanding the mechanisms involved in the PWV response to the solar activity then becomes overriding objective. The fact that the highly variable upper atmosphere is coupled to the middle atmosphere through chemical, radiative and dynamical mechanisms. Looking from a GPS perspective, analyses results of four geomagnetic storms and on a monthly basis in Antarctica show that solar related events variation was shown influence the recent atmospheric change through water vapor variability. The mechanism proposed is that when the intense of geophysical event from the Sun communicated via solar wind and solar flares that enter to the Earth's magnetosphere, their highest energy is transmitted directly to the upper atmosphere (the TEC change, such as follows the sunspots and geomagnetic activity), then downward to middle atmosphere through electric discharge from lightning-induced electron precipitation events (global electric circuit), collides the clouds modulated by cosmic-rays, whereby resulting the substantial alterations in chemical composition in the middle atmosphere and atmospheric change in the lower atmosphere. Temperature and incident density of energetic particles in the atmosphere are the main key in the interaction processes between the upper and the lower levels of the atmosphere may lead changes in precipitations and weather patterns. From long-term data, it can be also seen that variations in the solar EUV and TSI radiation has impact on the thermal structure and composition of the middle atmosphere. Variation in the cosmic ray and ozone concentrations, modulated by solar activity are also manifest in changes in atmospheric ionization, but details of their direct influence and its physical mechanisms is still needs to be further investigated. Advances in this field are necessitating a new brainstorm to establish the physical mechanisms of solar-climate relationship issues. Thus, the techniques introduced on analysis of the solar influence on PWV, are novel and believed to have new perspective and opportunity to explain of how solar activity exerts their influence on weather/climate changes.

REFERENCES

[1] IPCC. Summary for Policymakers. In: Climate Change 2007: The Physical Science Basis; Solomon, S.; Qin, D.; Manning, M.; Chen, Z.; Marquis, M.; Averyt, K. B.; Tignor, M.; Miller, H. L. Ed.; Cambridge University Press: Cambridge, United Kingdom and New York, NY, USA, 2007.

[2] Cucurull, L.; Navascues, B.; Ruffini, G.; Elosegui, P.; Rius, A.; Vila, J. J. Atmos. Oceanic Tech. 2000, 17, 773-787.

[3] SuomiNet. (2007). Real time integrated atmospheric water vapor and TEC from GPS. http://www.suominet.ucar.edu/intro.html.

[4] Stokes, G. M.; Schwartz, S. E. *Bul. Amer. Meteorol. Soc.* 1994, *75,* 1201-1221.

[5] Ware, R. H.; Fulker, D. W.; Stein, S. A.; Anderson, D.N.; Avery, S. K.; Clark, R. D.; Droegemeier, K. K.; Kuettner, J. P.; Minster, J.; Sorooshian, S. Earth Planets Space 2000, 52, 901–905.

[6] Jacob, D. Phys. Chem. Earth, Part A: Solid Earth and Geodesy 1995, 26, 523-527.

[7] Rocken, C.; Ware, R.; Van Hove, T.; Solheim, F.; Alber, C.; Johnson, J. J. Atmos. Oceanic Tech. 1995, *12*, 468-478.

[8] Bevis, M.; Businger, S; Chiswell, S. J. Appl. Meteorol.1994, 33, 379-386.

[9] Salonen, E.; Kahru, S.; Jokela, P.; Zhang, W.; Uppala, S.; Aulamo, S.; Sarkkula, S. Final report for ESA/ESTEC Contract 8025/88/NL/PR, 1990.

[10] Rocken, C.; Braun, J.; Meertens, C.; Ware, R.; Sokolocskiy, S.; VanHove, T. (1999). Water vapor tomography with low cost GPS receivers. http://www.gst.ucar.edu/gpsrg/arm98_paper.pdf.

[11] Suparta, W.; Mohd. Ali, M. A.; Yatim, B.; Fraser, J. G. J. Phys. Chem. Earth 2009, 34 (1-2), 72-87
[12] Alber, C.; Ware, R.; Rocken, C.; Solheim, F. Geophys. Res. Lett. 1997, 24, 1859-1862.
[13] Hoyt, D. V.; Schatten, K. H. J. Geophys. Res. 1993, 98, 18895-18906.
[14] Butler, C. J.; Johnston, D. J. J. Atmos. Sol.-Terr Phys. 1996, 58, 1657–1672.
[15] van Geel, B.; Raspopov, O. M.; Renssen, H.; van der Plicht, J.; Dergachev, V.A.; Meijer, H.A. J. Quat. Sci. Rev. 1999, 18, 331-338.
[16] Thejll, P.; Lassen, K. J. Atmos. Sol.-Terr Phys. 2000, 62, 1207–1213.
[17] Pittock, A. B. Rev. Geophys. Space Phys. 1978, 16, 400-420.
[18] White, W. B.; Lean, J.; Cayan, D. R.; Dettinger, M. D. J. Geophys. Res. 1997, 102, 3255–3266.
[19] Reid, G. C. Space Sci. Rev. 2000, 94, 1-11.
[20] Shindell, D.; Rind, D.; Balachandran, N.; Lean, J.; Lonergan, P. Science 1999, 284, 305-308.
[21] Marsh, N.; Svensmark, H. Phys. Rev. Lett. 2000, 85, 5004-5007.
[22] Kristjansson, J. E.; Staple, A.; Kristiansen, J.; Kaas, E. Geophys. Res. Lett. 2002, 29, 2107, doi:10.1029/2002GL015646.
[23] Gleisner, H.; Thejll, P. Geophys. Res. Lett. 2003, 30, 1711, doi:10.1029/2003GL017129.
[24] Perry, C. A. Intern J. Clim. 1994, 14, 969-983.
[25] Zhao, J.; Han, Y. B.; Li, Z. A. Chinese J. Astron. Astrophys. 2004, 4, 189-227.
[26] Fröhlich, C.; Lean, J. Geophys. Res. Lett. 1998, 25, 4377–4380.
[27] Suparta, W.; Yatim, B.; Mohd. Ali, M. A. Acta Geophys. 2010, 58(2), 374 - 391.
[28] Bromwich, H. D. Nature 1990, 343, 627–629.
[29] Connoley, W. M.; King, J. C. J. Geophys. Res. 1996, 101, 1335-1343.
[30] Suparta, W.; Abdul Rashid, Z. A.; Mohd. Ali, M.A.; Yatim, B.; Fraser, J. G. J. Atmos. Solar-Terr. Phys. 2008, 70, 1419-1447.
[31] Businger, S.; Chiswell, S. R.; Bevis, M.; Duan, J.; Anthes, R. A.; Rocken, C.; Ware, R. H.; Exner, M.; Van Hove, T.; Solheim, F. S. Bull. Amer. Meteorol. Soc. 1996, 77, 5-18.
[32] Rocken, C.; Hove, T. V.; Ware, R. *Geophys. Res. Lett.* 1997, 24, 3221-3224.
[33] Gutman, S. I.; Benjamin, S. G. GPS Solutions 2001, 4, 16-24.

[34] Ware, R. H.; Fulker, D. W.; Stein, S. A.; Anderson, D. N.; Avery, S. K.; Clark, R. D.; Droegemeier, K. K.; Kuettner, J. P.; Minster, J. B.; Sorooshian, S. J. Atmos. Solar-Terr. Phys. 2001, 63,1315-1330.
[35] Bevis, M.; Businger, S.; Herring, T. A.; Rocken, C.; Anthes, R. A.; Ware, R. H. J. Geophys. Res. 1992, 97, 15787−15801.
[36] Rocken, C.; Ware, R.; Van Hove, T.; Solheim, F.; Alber, C.; Johnson, J. *Geophys. Res. Lett.* 1993, *20*, 2631-2634.
[37] Duan, J.; Bevis, M.; Fang, P.; Bock, Y.; Chiswell, S.; Businger, S.; Rocken, C.; Solheim, F.; van Hove, T.; Ware, R.; McClusky, S.; Herring, T. A.; King, R.W. J. Appl. Meteorol. 1996, 35, 830-388.
[38] Gendt, G.; Dick, G.; Reigber, C.; Tomassini, M.; Liu, Y.; Ramatschi, M. J. Meteorol. Soc. Jpn. 2004, 82, 361-370.
[39] Friis-Christensen, E.; Lassen, K. Science 1991, 254, 698-700.
[40] Kaplan, E. Understanding GPS principles and applications. Norwood: Artech House, 1996.
[41] Spilker Jr, J. J.; Parkinson, B. W. Ed.; 1996. Overview of GPS operation and design, in Global Positioning System: Theory and Applications. Washington, D.C.: American Institute of Aeronautics and Astronautics.
[42] U. S. National Observatory (USNO). Current GPS constellation. http://tycho.usno.navy.mil/gps.html
[43] Dach, R.; Hugentobler, U.; Fridez, P.; Meindl, M. (Eds.). Bernese GPS Software Version 5.0; Astronomical Institute, University of Bern (AIUB); 2007.
[44] Hofmann-Wellenhof, B.; Lichtenegger, H.; Collins, J.; Atmospheric effect on the Global Positioning System: Theory and Practice; Springer Verlag, 2001; 5th revised edition, pp 99-117.
[45] Owens, J. C. Appl. Opt. 1967, 6, 51–58.
[46] World Meteorological Organization. WMO Technical Regulations 2000, 49, corrigendum.
[47] Thayer, G. D. *Radio Sci.* 1974, *9*, 803–807.
[48] Suparta, W. PhD Thesis; Faculty of Engineering, Universiti Kebangsaan Malaysia; 2008, pp 1-260.
[49] Davis, J. L.; Herring, T. A.; Shapiro, I. I.; Rogres, A. E. E.; Elgered, G. Radio Sci. 1985, 20, 1593-1607.
[50] Saastamoinen, J. *Bull. Geodesique* 1972, 106, 383-397.
[51] Niell, A. E. Earth Planets Space 2000, 52, 699–702.
[52] Foelsche, U.; Kirchengast, G. Phys. Chem. Earth, Part A: Solid Earth and Geod. 2001, 26, 153-157.
[53] Niell, A. E. J. Geophys. Res. 1996, 101, 3197-3246.

[54] Marini, J. W. Radio Sci. 1972, 7, 223– 231.
[55] UNAVCO. TEQC - The Toolkit for GPS/GLONASS/Galileo/SBAS Data. http://facility.unavco.org/software/teqc/teqc.html.
[56] SOPAC. http://sopac.ucsd.edu.
[57] Altamimi, Z.; Sillard, P.; Boucher, C. J. Geophys. Res. 2002, 107, 2214, doi:10.1029/2001JB000561.
[58] Suparta, W.; Mohd. Ali, M. A.; Yatim, B.; Misran, N. *IEEE Xplore* 2008, doi: 10.1109/SCORED.2007.4451402.
[59] Gendt, G. 1998. IGS combination of tropospheric estimates – experience from pilot project. Proc. IGS Analysis Centre Workshop, Darmstadt.
[60] National Weather Service, SWPC. http://www.swpc.noaa.gov/ftpmenu/warehouse.html.
[61] World Data Center for Geomagnetism, Kyoto. Geomagnetic Data Service, http://swdcwww.kugi.kyoto-u.ac.jp/wdc/Sec3.html.
[62] GSFC. http://cdaweb.gsfc.nasa.gov/cdaweb/istp_public/
[63] NOAA/NGDC. http://goes.ngdc.noaa.gov/data/
[64] Panasyuk, M. I.; Avdyushin, S. I.; Lazutin, L. L., et al. Cosmic Res. 2004, 42, 489-534.
[65] Veselovsky, I. S.; Panasyuk, M. I.; Avdyushin, S. I., et al. Cosmic Res. 2004, 42, 453-508.
[66] Chertok, I. M.; Fomichev, V. V.; Gnezdilov, A. A.; Gorgutsa, R. V.; Markeev, A. K.; Sobolev, D. E. Astron. Astrophys. Trans. 2005, 24, 45–52.
[67] Oler, C. Space Wea. 2004, 2, S08001, doi:10.1029/2004SW000076.
[68] Skoug, R. M.; Gosling J. T.; Steinberg J. T.; McComas J. T.; Smith, C. W.; Ness, N. F.; Hu, Q.; Burlaga, L. F. *J. Geophys. Res*. 2004, *109*, A09102, doi:10.1029/2004JA010494.
[69] Srivastava, N. *Ann. Geophys*. 2005, *23*, 2989-2995.
[70] Wang, Y.; Ye, P.; Zhou, G.; Wang, S.; Wang, S.; Yan, Y.; Wang, J. Sol. Phys. 2005, 226, 337–357.
[71] Tsurutani, B. T.; Mannucci, A. J.; Iijima, B.; Guarnieri, F. L.; Gonzalez, W. D.; Judge, D. L.; Gangopadhyay, P.; Pap. J. *Adv. Space Res*. 2006, *37*, 1583-1588.
[72] Gopalswamy, N.; Yashiro, S.; Michalek, G.; Xie, H.; Lepping, R. P.; Howard, R. A. Geophys. Res. Lett. 2005, 32, L12S09, doi:10.1029/2004GL021639.
[73] CELIAS/SOHO. http://umtof.umd.edu/sem

[74] Metatech. (2004). November 7-12, 2004 Geomagnetic Storm Summary. http://www.metatech-aps.com/Advisories.html

[75] Archive of Solar Terrestrial Activity Reports. http://www.dxlc.com/solar/old_reports/

[76] Rishbeth, H. Adv. Space Res. 1997, 20, 2149-2155.

[77] Cander, Lj. R.; Mihajlovic, S. J. Atmos. Solar-Terr. Phys. 2005, 67, 1118–1128.

[78] Warnant, R.; Pottiaux, E. Earth Planets Space 2000, 52, 1055-1060.

[79] Abdul Rashid, Z. A.; Momani, M. A.; Sulaiman, S.; Mohd Ali, M. A.; Yatim, B.; Fraser, G. J.; Sato, N. J. Atmos. Sol.-Terr Phys. 2006, 68, 1222-1236.

[80] Danilov, A. D.; Lastovicka, J. Intern. J. Geomagn. Aeron. 2001, 2, 209–194.

[81] **Lastovicka, J.** J. Atmos. Sol.-Terr Phys. 2002, 64, 697-705.

[82] GFZ. Index of Global Geomagnetic Activity. http://www-app3.gfz-potsdam.de/kp_index/qddescription.html.

[83] Mockler, S. B. (1995). Water vapor in the climate system. Special report, AGU, 2000 Florida Ave, N.W.; Washington, DC 20009, ISBN 0-87590-865-9, December 1995; http://www.agu.org/sci_soc/mockler.html.

[84] Gonzalez, W. D.; Joselyn, J. A.; Kamide, Y.; Kroehl, H.W.; Rostoker, G.; Tsurutani, B. T.; Vasyliunas, V.M. J. Geophys. Res. 1994, 99, 5771-5792.

[85] Parish, T. R.; Waight, K. T. Mont. weather rev. 1987, 115, 2214-2226.

[86] Nylen, T. H.; Fountain, A. G.; Doran, P. T. J. Geophys. Res. 2004, 109, D03114, doi:10.1029/2003JD003937.

[87] Vey, S.; Dietrich, R.; Jonhsen, K. P.; Miao, J.; Heygster, G. J. Meteorol. Soc. Jpn. 2004, 82, 259-267.

[88] Ataç, T. (2006). Private communication.

[89] Ataç, T.; Özgüç, A. Sol. Phys. 1996, 166, 201-208.

[90] van Lipzig N. P. M.; van Meijgaard, E.; Oerlemans, J. *J. Clim.* 2002, 15, 2758–2774.

[91] King, J. C.; Turner, J.. Antarctic meteorology and climatology; Cambridge University Press; Cambridge, 1997; pp 1-409.

[92] Gray, L. J.; Haigh, J. D.; Harrison, R. G. (2005). The influence of solar changes on the Earth's climate. Hadley Center Technical Note 62. Department for Environment, Food and Rural Affairs, UK.

[93] Fuller-Rowell, T. J.; Codrescu, M. V.; Moffett, R. J.; Quegan, S. J. Geophys. Res. 1994, 99, 3893-3914.

[94] Hathaway, R. H.; Wilson, R. M. Sol. Phys. 2004, 224, 5-19.
[95] Lyon, J. G. Science 2000, 288, 5473.
[96] Baker, D. N. J. Atmos. Solar-Terr. Phys. 2001, 63, 421-429.
[97] Dungey, J. W. Phys. Rev. Lett. 1961, 6, 47–48.
[98] Wright, A. (2004). The impact of solar weather. http://allrite.net/science/docs/ImpactOfSolarWeather.pdf.
[99] Langley, R. B.; Teunissen, P. J. G.; Kleusberg, A. Propagation of the GPS Signals. Springer-Verlag Berlin Heidelberg: GPS for Geodesy, 2nd Edition, 1998.
[100] Baker, D. N. (1986). Statistics analyses in the study of solar wind-magnetosphere coupling, in Kamide, Y.; Slavin, J. A. Ed. Solar-wind-Magnetosphere coupling. Tokyo: Terra Scientific Publishing Co.
[101] Johansson, J. M. (1997). Modeling of the earth atmosphere in space geodetic applications. In Geodetic application of GPS, Lecture Notes for Nordic Autumn School, number 16 in Report in Geodesy and geographical information system (GIS): 109-134, National land of survey, Sweden.
[102] Harrison, R. G.; Ambaum, M. P. H. Proceedings of the Royal Society A 2008, 464, 2561-2573.
[103] Burns, G. B.; Tinsley, B. A.; Frank-Kamenetsky, A. V.; Bering, E. A. J. Geophys. Res. 2007, 112, D04103, doi:10.1029/2006JD007246.
[104] Kunches, J. M.; Klobuchar, J. A. GPS Solutions 2000, 4, 77-78.

INDEX

N

O

P

R

S

T

U

V

W